EYES TO SEE

MICHAEL LAND

eyestosee

the astonishing variety of vision in nature

OXFORD

UNIVERSITY PRESS

OXFORD
UNIVERSITY PRESS

Great Clarendon Street, Oxford, OX2 6DP,
United Kingdom

Oxford University Press is a department of the University of Oxford.
It furthers the University's objective of excellence in research, scholarship,
and education by publishing worldwide. Oxford is a registered trade mark of
Oxford University Press in the UK and in certain other countries

First Edition published in 2018
Impression: 1

Published in the United States of America by Oxford University Press
198 Madison Avenue, New York, NY 10016, United States of America

British Library Cataloguing in Publication Data
Data available

Library of Congress Control Number: 2018944716

ISBN 978–0–19–874771–0

Printed in Great Britain by
Clays Ltd, Elcograf S.p.A.

PREFACE

The study of vision spans most divisions of science. Physics is involved in the optics of image formation, and chemistry in the way light energy is converted to nerve activity. The biology of animal vision is a vast subject, as different animals use sight for many different purposes, from simple navigation to recognizing individuals. The nature of human vision has attracted philosophers since antiquity, and modern psychologists have explored the processes of perception through experiment, and more recently, by techniques such as brain scans. There is much more to find out—we still don't even know how to ask the question 'Why does red look red?'

In fifty years working on vision, I have had contact with, and sometimes immersion in, many aspects of the subject. I began by studying some of the more unusual eyes of invertebrates, and after a meandering journey through the animal kingdom, found myself working on the ways that people use eye movements to find the information they need to do things. I suppose that at different times, I have been an applied physicist, a physiologist, an ethologist, and a psychologist, but I did manage to avoid most chemistry. I'm happy to admit that this diversity of interest was no accident. At school my best subjects were physics and biology (these coincided with the best teachers). My main hobby was photography. It seems that the child was indeed the father of the man.

In this book I have selected those aspects of eyes and visual systems that have intrigued me most, in the hope that I can convey some of the joy that I have had in the study of them. This is not just my story: there have been many others—colleagues, students, and people I have never met—that have contributed to the milieu in which I worked, and I have drawn them into the narrative. I have always studied the things that seemed most interesting to me at the time, so my progression in the field has been opportunistic rather than logical. However, in the book, I have tried to organize the chapters into related themes, to provide a reasonably coherent framework.

It is a pleasure to thank the staff at Oxford University Press and in particular, my editor Latha Menon. She suggested that I write the book, she has read and greatly improved the text, and has guided its progress through to publication.

CONTENTS

1. Early Eyes   1

2. Compound Eyes and Insect Vision   27

3. Vision in the Ocean   61

4. Establishing Identity   92

5. Where Do People Look?   127

6. The Mind's Eye   148

7. The Evolution of Vision   171

Endnotes   179
Glossary   183
Art Credits   188
Index   195

1

EARLY EYES

I am going to begin this book, unashamedly, by writing about my first proper scientific discovery, which I still regard as an extraordinary piece of good luck. As a beginning postgraduate student at University College London, I had managed to persuade John Gray, my supervisor, that scallops would be a good animal to study. Unlikely as this seems, it was more than just a whim. Around the inner margin of a scallop shell is the mantle—the bit you don't see at the fish shop—and this contains a mass of sensory tentacles, embedded in which are about sixty rather pretty eyes, each about a millimetre across (Fig. 1, Plate 1). At the time it was not the eyes themselves that attracted me, but the fact that there were a lot of them, feeding into a single central ganglion attached to the large and very edible muscle. Why so many eyes? What were they telling the brain? What use did the brain make of this information in organizing the animal's behaviour? There was clearly a puzzle here. My idea was to record the electrical activity in the nerves from the eyes to the ganglion, record the output to the muscle, and then record from the cells in the ganglion to work out what processes triggered a response. Scallops don't do a lot: their main behaviour is to shut when things approach.

John Gray was a sensory physiologist who worked on vertebrate touch receptors, but he liked the scallop idea and gave me much advice and help, though he left part way through my Ph.D. to become secretary of the Medical Research Council. This resulted in a heart attack and a knighthood, and he retired to Plymouth, where he continued to work on fish hearing for another 30 years.

Before getting too deeply involved in the dark arts of electrophysiology, I had to work out how the scallop's eyes worked. There were plenty of existing descriptions, and from these one could conclude that scallop eyes were

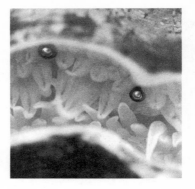

Fig. 1. Two eyes of the scallop *Pecten maximus* looking out between the tentacles of the mantle edge. The eyes are a millimetre across, and the images of some of the lights in the room are visible in the pupils.

conventional, with more or less the same layout as a fish eye, with a lens, a retina, and a reflecting mirror-like layer behind the retina, much as in the eye of a cat (Fig. 2). One day in 1964, I was looking into one of these eyes with a dissecting microscope, and noticed that there was an upside-down image of me in the eye (Fig. 2). I waved my hand and the image mirrored my action. I didn't initially think much of this, because, after all, eyes are supposed to contain images. But after a while it dawned on me that something was wrong. If you look into a human eye, you can't see an image within it, because you are trying to see the image through the same lens that formed it. What this means is that the image on the human retina, seen from outside through the lens, is back where the object was—somewhere a long distance outside the eye, and very definitely not in it. (You can see the image in the human eye with an ophthalmoscope, but that instrument contains a lens system that neutralizes the power of the eye's optics, allowing the retina to be viewed directly.)

In the scallop eye, the image was very definitely *in* the eye, and not outside it. Another curious feature was that the image was very bright, and the only structure that could account for this brightness was the mirror behind the retina. When all this had sunk in and stirred around for a while, I realized that the image that I could see in the scallop eye was formed not by the lens, but by the mirror itself.[1] Like convex lenses, concave mirrors form images: a concave shaving mirror will throw a real image of a window onto a wall.

2

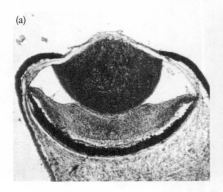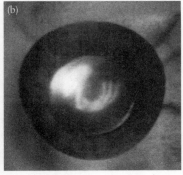

Fig. 2. a) Undistorted frozen section of a scallop eye showing the lens, the thick retina, and the mirror beneath it above the pigment layer. b) Image of the author's hand holding the microscope lens used to take the photograph.

The optical system of the scallop's eye was based not on a lens but on a concave mirror, and this was a system not previously known in biology. Newton had first used this principle in his reflecting telescope of 1668.

This was quite something. The marine biologist Eric Denton once remarked to me, when we were working together on fish scales: 'When you get a good result, go and have a good dinner; then at least you've had a good dinner.' I did indeed celebrate, and fortunately, scallops still had images in their eyes the next day. There was much more to do though. For a start each eye did have a lens, so there was more to the optical system than just the mirror. The lens turned out to be soft and with a low refractive index, so it did not add much to the power of the mirror. On its own it would have formed an image a long way behind the back of the eye. It did, however, have an odd profile to its front surface, and tracing rays through a section of the lens suggested that its unusual shape might improve the optical performance of the mirror. (For astronomers, there is a connexion here with the Schmidt corrector plate used in reflecting telescopes.)

Then there was the retina itself (Fig. 3). This is a thick structure, with two layers, each having a different kind of photoreceptor. These were referred to in the literature as the proximal retina (nearest to the mirror) and distal retina (furthest from the mirror and in contact with the back of the lens). Each

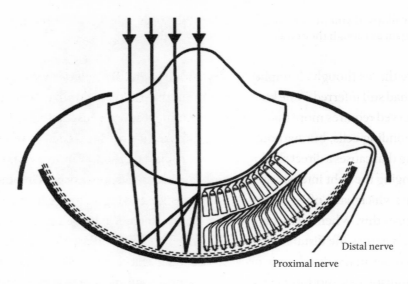

Fig. 3. Optical diagram of the scallop eye, showing ray paths on the left and the two retinas on the right, each with its own optic nerve.

retina has its own nerve, and the activity in these nerves in response to light had been recorded by H.K. Hartline as early as 1938. What Hartline had found was that the proximal retina responded with action potentials[2] when the light went on, and the distal retina when the light went off. The obvious question this now raised was which retinal layer receives the image, because the separation of the receptive regions is too great for the image to be seen by both. The image produced by a concave mirror, according to basic laws of optics, lies halfway between the mirror surface and its centre of curvature. In the scallop's eye, this corresponds exactly with the receptive regions of the cells of the distal retina. The proximal retina, with its receptive regions touching the mirror, has no image.

When a scallop detects the movement of objects nearby, it closes its shells so that the eyes and tentacles are protected from creatures that might want to nibble them. This behaviour is well known to divers, and amazingly, it was known to Aristotle in the fourth century BC. I can't resist this quote from his *History of Animals*.[3]

4

Scallops, if you present your finger near their open valves, close them tight again as though they could see what you were doing.

I like the 'as though': it implies that Aristotle hadn't actually seen the eyes, but had still inferred some kind of vision. For the scallop to do what Aristotle observed requires more than a simple dimming response. Just having 'off'— responding cells, but without an image, makes it possible to detect something that causes a direct shadow, but not an object that simply moves without changing the light intensity at the eye. This shadow response occurs in other clams which have 'off'—receptors, but no optics. Having an image in the eye changes this. When a dark object moves in the water near a scallop, its image travels across the distal retina, and as it does so it sets off a series of action potentials in the distal optic nerve as it darkens successive cells. This is not a sophisticated motion detector of the kind found in insect or vertebrate eyes, where brightness changes across the retina are compared in time and space; here each cell just responds to a decrease in light, but that is all that is needed. As my colleague Dan-Eric Nilsson puts it, this is basically a burglar alarm: it doesn't need to register speed or direction, just local dimming.[4]

There are still some mysteries. Each retina has about 5000 receptors, and although the distal retina now has a useful function, this doesn't seem to be true of the proximal retina, which has no image. Scallops do swim, towards or away from rocks and other landmarks, and the distal retina, with its very short-lasting responses, doesn't seem suited to this. Presumably the proximal retina, which gives sustained responses to light changes, does this job in a blurry sort of way. In *Pecten maximus*, the species I worked on, there are about sixty eyes looking in all directions, so the proximal retinas between them should be able to work out in low resolution the distribution of light around the animal. The sheer number of eyes also means there is huge overlap between the roughly 90° fields of view of adjacent retinas. If an eye is looking in one direction, the next eye will be pointing only 6° away, which means that the distal retinas of up to thirty eyes will, in principle, all see the same moving object. Perhaps this overlap simply reflects the need for a stronger collective signal when light is dim. On the other hand, perhaps resolution is best only in the centre of the eye, as suggested by the profile of the lens, so that for small distant movements the effective field of view of each eye could be quite small. In any event, scallops can detect movements as small as 2°, the angular width of a single distal receptor projected into the space around the animal, so the acuity of the eye is as good as it reasonably can be.

The two kinds of cells in the retina turned out to be of great interest in terms of the evolution of eyes in general, although neither I nor anyone else realized this at the time. Electron microscopy, particularly that of Vernon Barber at University College London (UCL), had shown that the proximal (on) receptors had a conventional 'rhabdomeric' structure, in which the molecules that respond to light (rhodopsin) are held in the membrane of short projections known as microvilli (Fig. 4). This is the kind of receptor usually found in other molluscs such as snails and squid, and in insects and other arthropods. The distal (off) receptors are of a different 'ciliary' type. Here the photopigment molecules are held in the membranes associated with one or more cilia (Fig. 4).

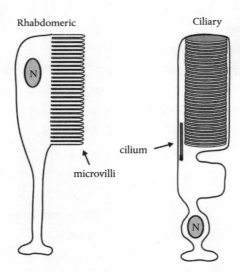

Fig. 4. Rhabdomeric and ciliary receptors showing the expanded membranes that carry the rhodopsin molecules. These examples are not from a scallop: the rhabdomeric example comes from a fly, and the ciliary example from a vertebrate rod, but the two receptor types are found in various versions throughout the animal kingdom. N is the cell nucleus.

Cilia are structures with a characteristic internal skeleton made of narrow tubules. In other organs they are usually mobile structures involved in the movement of fluid. In sensory cells they no longer have this ability to move, but are concerned—in some less then obvious way—with the extraction of sensory information. As we shall see later, these two receptor types are primordial, going back to a time in the Precambrian before the major animal phyla separated from each other. Presumably scallops inherited these cell types from distant relatives. The bivalve molluscs certainly go back to the Cambrian

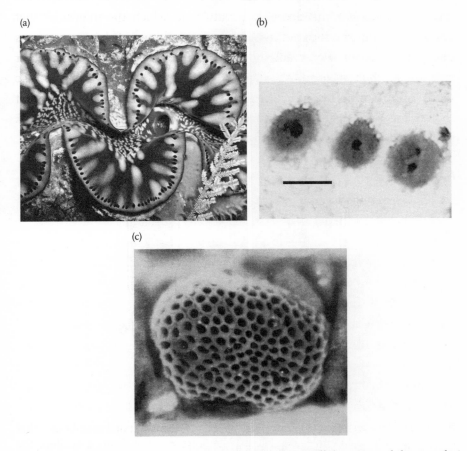

(a)

(b)

(c)

Fig. 5. a) Eyes of the giant clam *Tridacna*, seen here as small dots around the mantle. b) Close-up of two eyes showing the 'pinholes'. Scalebar is 0.5 mm. c) The compound eye of the arc clam *Barbatia*.

Period (540–485 million years ago), although the scallops themselves are a relatively new family, arising about 250 Mya, just after the mass extinction at the end of the Permian Period (see Fig. 6). What makes scallop eyes special is that it is very unusual to find both receptor types in the same eye.

Other bivalve molluscs have come up with different solutions to the problem of combining off-responding receptors with optics of some kind, to help them to detect predators at a distance. Pacific giant clams (*Tridacna* species) have huge numbers of eyes around the mantle. These are pinhole eyes, with neither lenses nor mirrors.[5] Each has a simple pigmented aperture just a tenth of a millimetre wide with some receptors behind it (Fig. 5). Their resolution is about 16°, which means they can just about detect a 10-cm fish at about 40 cm; this should give the clams enough time to close and protect their vulnerable tissues. In another group of bivalves, the arc shells, small compound eyes have evolved for the same purpose. In *Barbatia* each of the 300 or so eyes consists of a convex array of lens-less pigmented tubes each with a receptor at the bottom, and pointing in a slightly different direction from its neighbour (Fig. 5). Each eye has up to 130 of these detectors.[4] Although as compound eyes (see Fig. 8), they are nothing like as sophisticated or acute as an insect eye—to which they are completely unrelated—the clams are nevertheless capable of responding to a moving stripe only 6° wide, which is rather less than the width of a hand at arm's length. Outside the molluscs, independently evolved compound eyes are also found in annelid tube worms, which have small eyes, similar to those of arc clams, on the ends of their feeding tentacles. Again, the function is to allow the animal to withdraw its tentacles back into its tube, if danger threatens. In none of these animals is there any suggestion that the eyes are doing anything other than acting as simple movement detectors, although in the case of scallops they may also provide limited information for guiding swimming. This contrasts with the eyes of insects, which allow the animal to control flight, recognize food plants, navigate by landmarks or cues in the sky, and so on. It is not so much the optical quality of the eye that determines its role in behaviour, but the capacity of the brain to make use of the information the eye supplies. Bivalves are not very bright.

The Evolutionary History of Eyes

It takes a real effort of imagination to comprehend the huge time scale of animal evolution. The 800 million years since multicellular animals first began is two hundred thousand times longer than any written record of human activity. As with history, the only way to cope with vast time scales is to break them up into units that are more manageable. It may be the 50-year reign of a historical monarch, or, on a geological scale, the roughly 50 million years between the events, often cataclysmic, that produced changes in the rock patterns of the Earth's crust. These intervals make up the eras of geological and evolutionary time (Fig. 6).

The animals living in the world today are the tips of the branches of an evolutionary tree that has been ramifying since its dim origins in the Precambrian seas. The family structure of the animal kingdom, with phyla, classes, families, genera, and species, so ably described by Linnaeus in the eighteenth century, are the trunks, branches, and ultimately twigs of this tree. Its growth has not been straightforward, and many branches have fallen during its long history. The trilobites, for example, were successful scavengers of the sea floor with a great many distinct species. They lasted for 300 million years, but then died out in the great Permian extinction event, 250 million years ago, which wiped out 96% of all marine species. To put this time span in perspective, hominids have only been around for about 2 million years, and there is little reason to suppose that we will share

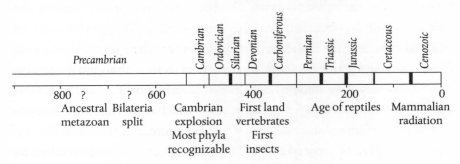

Fig. 6. Timeline of evolution, showing the geological periods, the timescale in millions of years before the present era (mya), and the main events in the evolution of animals.

the trilobites' longevity. The species on the Earth today are just a cross-section of a complex and unfolding pattern, and a comparable cross-section taken at any time in the last half-a-billion years would have produced almost as complex a pattern of species—though hardly ever the same ones as those alive today.

The First Receptors

After my time at UCL in 1967 looking at scallop eyes, I went to Berkeley on a post-doctoral fellowship. Berkeley at the time still had the reputation as 'the Athens of the West' and had a friendly, hippyish, feel to it. The site was beautiful, with groves of redwoods, and glimpses of San Francisco and the Bay. The smell of the redwoods and bay trees, and the eucalyptus on the hills above, made it a spirit-lifting place to work. Sadly, this gentle mood changed during the four years I was there. A combination of the Vietnam War, underfunding by a state government hostile to education, and a hardening drug culture led to increasing strife, and eventually to riots serious enough to involve the National Guard—equipped with tanks. Life went on, but the idyll was over.

My sponsors in Berkeley were Gerald Westheimer, whose field was human vision, especially eye movements, and Richard (Dick) Eakin, a zoologist who studied the anatomy of animal photoreceptors. Both had a great effect on the development of my scientific thinking. Gerald gave me all sorts of valuable ideas, and in later chapters, I will develop these, but it is Dick Eakin's work that is relevant here.

Eakin came to Berkeley as an undergraduate in 1929, joined the faculty in 1940 and built up the Zoology department. He was a man of many parts, literally. Noticing that students in his Zoology class found some of the course-work dull, he began a series of lectures in which he impersonated famous scientists, in full costume and make up. He began with William Harvey, complete with ox heart and tomato juice. He was a gifted actor—he would say ham— and these shows became special events, drawing audiences from across the campus. Over the years, he built up a repertoire including Gregor

Mendel, Charles Darwin, Louis Pasteur (Fig. 7), William Beaumont, and the embryologist Hans Spemann. The lectures were published by the University of California Press in 1975, with splendid pictures of the impersonations.[6] Eakin's innovative teaching brought national awards, as well as making the pages of *Life* magazine and *Der Spiegel*. I should add that Eakin was a lifelong Christian, as well as an evolutionary biologist: a combination not particularly common in the USA. He died in 1999 at the age of 89.

Eakin spent much of his research career as an electron microscopist, working on the fine structure of photoreceptors in the eyes of just about everything. By 1963 he had assembled enough information to classify receptors as either rhabdomeric or ciliary, depending on whether their photoreceptive membranes were based on microvilli or cilia (Fig. 4). He went further, and to explain this I need to delve briefly into some mysteries of embryology. The phyla that make up most of the animal kingdom separate into two main divisions: the protostomes and deuterostomes. These words mean 'first mouth' and 'second mouth', from the way the mouth and anus develop in the early embryo. The protostomes develop the mouth first. This distinction has been regarded as evolutionarily ancient and fundamental. It separates the arthropods, molluscs, and worms (and others), which are the protostomes, from the starfish, sea urchins, and chordates (from which the vertebrates are derived), which are the deuterostomes. Eakin had concluded that the different receptor types map nicely onto this division, with the protostomes having rhabdomeric receptors, and the deuterostomes having ciliary receptors (Fig. 7). The implication would seem to be that the two types of receptor arose in the Precambrian, at much the same time as the two great embryological divisions of animals split from each other. By and large, Eakin's classification worked well. The photoreceptors of arthropods (crustaceans, insects, and spiders) are all rhabdomeric, as are the receptors of squids, snails, and worms. The rods and cones in the eyes of all vertebrates are based on cilia, as are photoreceptors in most other chordates, although starfish photoreceptors have a somewhat intermediate structure. But there were problems with Eakin's simple division. Not least of these were the eyes of scallops. Scallops are undoubted protostomes, but they have both

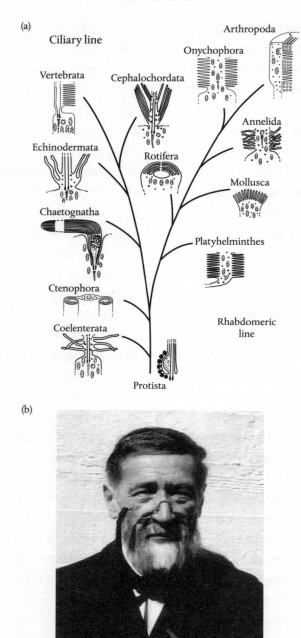

Fig. 7. a) Eakin's ciliary and rhabdomeric receptor classification. b) Richard Eakin as Pasteur.

rhabdomeric and ciliary receptors in the same eye, and this certainly doesn't fit Eakin's scheme. In fact there were sufficient other exceptions to the linking of protostomes with rhabdomeric receptors and deuterostomes with ciliary receptors that the whole idea came to be questioned. And the implications of the scheme—that the two types of receptors arose separately in the two great animal divisions—could not be sustained. As we shall see, answers emerged many years later, when the structures of the photoreceptor molecules themselves—the opsins—were worked out, and the biochemistry of the chains of events linking light absorption to the electrical response of the receptors, called the transduction cascade, became clear. But Eakin's electron microscopy continued to provide the structural groundwork for all the later discoveries.

The photoreceptor cells that respond to light, and make vision of any kind possible, predate eyes themselves by several hundred million years. Eyes that we would recognize as eyes arose during the Cambrian Period, beginning about 540 Mya (Fig. 6). Before that there were undoubtedly small eyecup-like structures containing photoreceptors, but we know little about them because they left no fossils. However, we do know, from their presence in animals that go back to Precambrian times, that both types of photoreceptor were already present. What, then, can we infer about their origins? From the extent of genetic overlap between protostomes and deuterostomes, it is clear that they had a common ancestor, sometime before 600 Mya. This ancestral animal line is usually referred to as the Bilateria or Urbilateria because these creatures would have had the form of 'proper animals': bilaterally symmetrical and with a head at the front. This distinguishes them from other ancient phyla such as the cnidarians (or coelenterates)—jellyfish, anemones, and hydrozoans—which are usually radially symmetrical, and the sponges, which do not have much recognizable symmetry. Further back still, possibly before 800 Mya, these multicellular animals (the Metazoa) evolved from single-celled creatures. The question is this: were the earliest phyla equipped with photoreceptors, and if so were there two types, or just one from which the other was derived? Since many cnidarians have photoreceptors, and these predate the Bilateria, we can infer that photoreceptors were probably present in some form right from the beginnings of the Metazoa.

In the early 2000s, two remarkable discoveries were made by Detlev Arendt and his colleagues in Heidelberg. The first was that the ragworm *Platynereis*, a protostome with rhabdomeric receptors in its eyes, also has ciliary receptors in its brain. The second, even more surprising, was that mice and fish, whose rods and cones are ciliary, contain light-sensitive ganglion cells in their retinas whose visual pigment is closely related to the rhabdomeric photopigments of invertebrate protostomes, and associated biochemistry more like that of an insect than of vertebrate rods and cones. In other words, it seems that some protostomes can have both rhabdomeric and ciliary photoreceptors, and so can the deuterostome vertebrates.[7] This is the same message as the scallop's eye gave us, but now backed up with some heavy-duty molecular biology. The conclusion is clear: both types of receptor must predate the protostome-deuterostome split, and both originated early in the evolution of the bilaterians. So there is no one-to-one correspondence between photoreceptor type and embryological superphylum, as Eakin had envisaged. It seems that the basic components for making receptors of either type are present in the genomes of all bilaterians. Nevertheless, which receptor type makes up the retina in the eyes themselves does seem to be determined mainly by which lineage they come from: rhabdomeric in protostomes, and ciliary in deuterostomes. One might think that the receptors in the eyes of cnidarians, which separated from the bilaterians even earlier in metazoan evolution, would throw some light on the question of which type of receptors came first. However, cnidarian photoreceptors are intermediate, with both cilia and microvilli, and a photopigment different from either of the principal bilaterian types. Biochemically they have much in common with both, so presumably they too share an even earlier common origin.

A final word on the receptors of scallop eyes. By and large the protein components of the photopigments of protostome and deuterostome eyes—the opsins—form two families: r-opsins and c-opsins (i.e. rhabdomeric and ciliary). Within each family the opsins are similar, but the similarity between the families, in terms of the structure of their genes, is low. The opsins of the ciliary receptors of the mantle eyes of scallops, and other bivalve molluscs, are different again. These so-called Go-opsins have little in common, genetically, with either of the other families. The scallop transduction cascade is similar to that of other ciliary photoreceptors, but it is still not quite the same. Most of the eyes of bilaterians of both embryological types are cephalic—that is, they are borne on the heads—but the eyes on the mantles of bivalves are not formed from head tissue, and they almost certainly have a different evolutionary origin. What we seem to have is a new type of receptor making use of some of the components, or genetic modules, available in the bilaterian genome for making eyes, but exploiting them in a way rather different from either of the 'standard' versions found in cephalic eyes.

The First Eyes

The principal types of photoreceptor, then, go back deep into the Precambrian, certainly to the early bilaterians, and possibly earlier. But eyes themselves, or at least organs that we would have no difficulty recognizing as eyes, do not appear until much later. Most of the eye types that we still see in animals alive today first appeared in the Cambrian Period, which began on 541 Mya and lasted about 65 million years (Fig. 6). This was a period in which a great increase in animal profusion and diversity occurred—the so-called Cambrian Explosion. There is no real consensus on what made this happen, but whatever the cause it had many consequences for animal structure. The results of these changes have been beautifully preserved in soft rocks in several locations, the most famous of which is the Burgess Shale of Western Canada.[8] What these fossils reveal is that animals became larger, more mobile, more heavily armoured, and evolved much better eyes than they had before. One factor which seems to link all these changes is the development of active predation, presumably driven by the fact that eating meat is nutritionally more efficient than rasping algae from rocks. To find and catch prey, predatory animals need to be fast and to have good eyesight. Equally, prey animals need good eyesight to see predators at a distance and to evade them. So a sort of visual arms race began, resulting in an increase in eye size, and with it much better image resolution.

Among the first of these Cambrian animals were the trilobites. These are arthropods with jointed limbs, distant relatives of the modern chelicerates, the group that includes horseshoe crabs, scorpions, and spiders. The trilobites had large compound eyes, and we are fortunate that the material they used to make their lenses was calcite, a hard mineral that survives indefinitely. They effectively came pre-fossilized, so we have an excellent record of at least the surface features of these eyes. Other arthropods with compound eyes were abundant in the Cambrian fauna, and these included the gigantic *Anomalocaris*, a predator up to two metres long with 16,000 facets in each eye, comparable with the eyes of modern dragonflies. More typical crustaceans start to appear by the end of the Cambrian. There were also representatives of other modern phyla, particularly the chordates and the molluscs, although we have much less information about their eyes, because they were soft, and if they fossilized at all they usually only left dark imprints with little detailed structure. Nevertheless, some chordates such as *Metaspriggina* had prominent camera-type eyes, and this chordate has been described by Simon Conway Morris, its co-discoverer, as a 'primitive fish'. Jawless fish that still have living relatives, the lampreys, arose a little later in the Ordovician. Lampreys have eyes remarkably similar to all other fish, so we know that the basic vertebrate eye plan was already well established by about 450 Mya, and presumably came into being during the Cambrian. Some early cephalopod molluscs from the Cambrian, such as *Nectocaris*, also had camera-type eyes, in this case on stalks. It seems from this that most of the major eye types still present evolved to quite a high level of sophistication more than 500 million years ago.

At this point it seems helpful to set out a catalogue of the kinds of eye present in the animal world today. Figure 8 illustrates the eight different known basic types of eye.[9] They divide into two broad groups: single-chambered eyes, also known as simple or camera-type eyes (a, c, d, g), and compound eyes (b, e, f, h). These have fundamentally different geometries; in the single-chambered eyes, the receptors line a single concave pit, but in compound

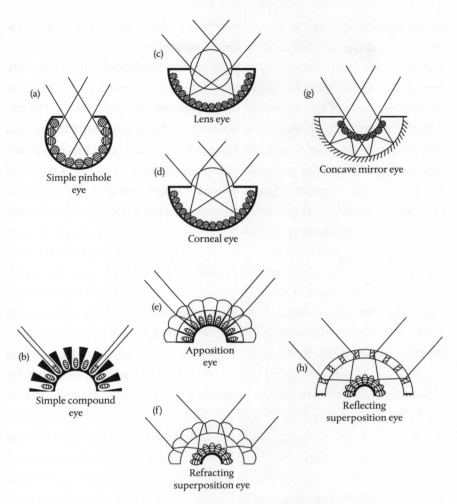

Fig. 8. The eight principal types of eye known in animals.

eyes, the receptors are in separate pigment tubes arranged around a convex dome. Both kinds of eye have evolved many times and in several different phyla, and were almost certainly present in their simplest forms (a and b) before the Cambrian. In terms of the optics involved, they produce images in one of three different ways: simple shadowing (a and b), lenses (c–f), or mirrors (g and h). We cannot be sure exactly when all the different variations evolved, because that requires a knowledge of soft tissues that do not fossilize. However, the only eye type that definitely evolved *after* the Cambrian is our kind of eye (d), in which most of the optical power comes from a cornea separating air from fluid. Such eyes will only work on land, and the first amphibians did not appear until the late Devonian, more than 100 million years after the end of the Cambrian. Among the compound eyes, (b) involves only shadowing and is still found in some bivalve molluscs (*Barbatia*, Fig. 5). Adding lenses produces *apposition* eyes (e), the most common type of compound eye, and rather more complex optics involving either lenses (f) or mirrors (h), and it gave rise to *superposition* eyes in which much of the eye surface contributes to a real image. We will meet all these eyes again in later chapters.

Early Vision

It is a long way from the first rhodopsin molecules of the early metazoans to the complex high-resolution eyes of humans or squid. The enormity of this gulf famously worried Darwin:[10]

> To suppose that the eye with all its inimitable contrivances for adjusting the focus to different distances, for admitting different amounts of light, and for the correction of spherical and chromatic aberration, could have been formed by natural selection, seems, I freely confess, absurd in the highest degree.

But he then argues that:

> ...if numerous gradations from a simple and imperfect eye to one complex and perfect can be shown to exist, each grade being useful to its possessor...then the difficulty in believing that a perfect and complex eye could be formed by natural selection, though insuperable to our Imagination, should not be considered as subversive to the theory.

The crucial point here is that whatever intermediate visual structures were produced during the evolution of eyes, they all had to work—evolution can only proceed by improving on what is already up and running, and by providing a survival advantage. The task, then, for anyone trying to produce an account of vision and visual structures before the evolution of modern eyes in the Cambrian, is to come up with a plausible sequence of advances, each building on the last. None of this could happen in a single step; receptors had to be provided with appropriate transducer cascades; optical systems had to evolve from appropriate materials; brains had to evolve that could make use of the new images, and so on. Fortunately, there are many examples of animals still present today that retain the characteristics of some of the earlier stages in this progression. When a species becomes well suited to its environment, and that environment doesn't change, evolutionary pressures subside, and little or no changes occur over long periods of time. For example, the lamp-shell *Lingula*, which lives in intertidal sand and mud, has remained virtually unchanged since the Cambrian. The similarities between many of the marine larvae of molluscs and annelid worms (trochophore larvae) implies that they too have changed little since the protostomes split into separate phyla in Precambrian times. These larvae carry a variety of small visual structures, often no more than one or two photoreceptors backed by pigment, and these are exactly the kinds of transitional structures one would expect to find in the early stages of visual evolution. It is worth remembering that there are only three major animal groups—the arthropods, the cephalopod molluscs, and the vertebrates—in which eyes and visual systems with a full range of capabilities evolved (Fig. 9). In all others the process of visual evolution did not go beyond a stage that is less than what Darwin would have described as 'perfect and complex'.

In 2009 my colleague Dan-Eric Nilsson, from Lund University in Sweden, proposed a scheme which links advances in the structure of photoreceptors and eyes to advances in the kinds of behaviour that such structures can provide (Fig. 10). I give his analysis here because it makes beautiful evolutionary sense. According to this scheme visual evolution proceeded in four stages, each boosted by a particular innovation.[11]

Stage 1 consists of non-directional photoreception, provided by cells containing photopigment linked to enough transduction machinery for the organism to move in response to slow changes in light intensity. Such

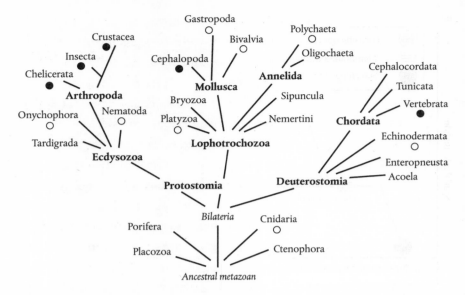

Fig. 9. The main divisions of the animal kingdom, indicating the levels of eye specialization found in each branch. The protostome/deuterostome split occurred early in the evolution of the bilaterians. (I use the anglicized forms of the Latin names here and in the text.) Thereafter the protostomes are divided into the jointed-limbed arthropods, the molluscs, and the annelid worms, as well as several other minor phyla. The deuterostomes are divided into the echinoderms (starfish relatives) and the chordates, which gave rise to the vertebrates and, ultimately, ourselves. Eyes providing spatial vision of low resolution are present in many phyla (open circles), but high-resolution multipurpose eyes are only found in the arthropods, the cephalopod molluscs, and the vertebrates (filled circles). The open and filled circles correspond to Nilsson's stages three and four in Figure 10.

receptors can be surprisingly useful. Light intensity can be a proxy for measuring depth in the water column, or nearness to the surface for an animal living in sand. They can be part of a clock for setting a daily activity cycle, and can provide a protective mechanism for avoiding the harmful effects of short-wavelength light. They may even provide a simple way of selecting the right light environment by simply causing the animal to go slower when the conditions are right, and faster when they are not. This is known as 'orthokinesis' in the older literature. Receptors such as these were the visual organs of the earliest metazoans.

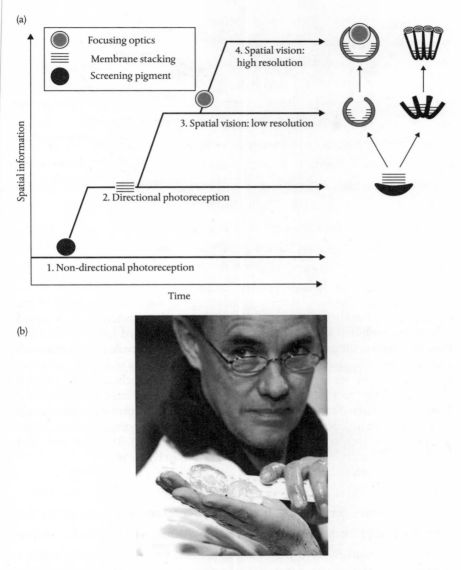

Fig. 10. a) scheme illustrating four stages in the evolution of visual systems, from non-directional photoreception to high-resolution spatial vision as found in modern vertebrates and arthropods. The insert shows the types of eye involved at each stage. b) Dan-Eric Nilsson holding the lenses of the eyes of a giant squid.

In Stage 2 photoreceptors can become directional, by being combined with cells containing a dark screening pigment, such as melanin, but on one side only. This means that the cell or cells have a partially restricted field of view, which in turn means that they can direct their possessor towards or away from light (often referred to as phototaxis). Directional photoreceptors can also provide a way of telling up from down, and so allow an animal to keep the right posture in a water column. They can also provide a limited protection from predators that cast shadows. Such simple receptors are common in planktonic larvae, such as the trochophore larvae already mentioned. Some flatworms and nematodes have similar receptors, usually in pairs. Another universal advance is in the construction of the photoreceptors. These now have a greatly increased area of membrane, and a correspondingly larger number of photoreceptor molecules. This means they capture much more light, certainly than a Stage 1 receptor, which allows them to operate in dimmer environments. Probably more important, they can signal changes much faster; this is because they don't have to count photons for so long before they can reliably signal a change in intensity. Darwin considered a simple combination of receptor and pigment cell as the first structure he would recognize as an eye, but Nilsson and I prefer to keep the term *eye* for the next stage, where each structure provides some degree of directional resolution.

Stage 3 structures begin to look like eyes. Typically they have a number of receptors in a small cup of dark pigment (Fig. 8a). This means that receptors have different fields of view—those in the left of the cup have a view, through the aperture, to the right, and conversely, for receptors on the other side of the cup. The point here is that there is now some resolution of direction within the eye itself—the beginnings of what we can call spatial vision. Without lenses the resolution will be awful. A receptor might have a field of view of 45° or more, but this will be enough to guide the animal away from or towards a large boulder or rocky outcrop that might provide shelter. These eyes can also be used to determine the direction of major celestial objects such as the sun and moon, and can probably be used to track an animal's progress through the environment. Eyes like this are very common: they are typical of flatworms, molluscs such as limpets, annelid worms, and again in many marine larvae. We have met already proto-compound eye versions of low-resolution eyes (Fig. 8b), as predator detectors in bivalve molluscs. Here, rather than having many receptors in a single cup, each receptor has its own screened tube.

So far, we can assume that all the structures just mentioned were present in at least some Precambrian bilaterian animals. The next level, Stage 4, in which much-improved optics led to high-resolution vision in animals with both single-chambered and compound eyes, did not come about until the Cambrian Period. Two inventions had to go hand in hand: optical systems that provided well-resolved images, and brains capable of handling the greatly increased quantity of visual information. To take the brains first, in animals that use vision for one or a few purposes—predation in a simpli-fied environment like the sea, or predator avoidance or simple guidance by the sun—the parts of the brain devoted to vision are usually small and hard to distinguish from the rest of the nervous system. However, in the three classes that use vision for multiple purposes—the vertebrates, the arthro-pods, and the cephalopod molluscs—roughly half the volume of the brain is devoted to vision (Fig. 11). A good eye, allied to a good brain, can be put to any number of uses, for instance, control of locomotion (including flight); navigation using landmarks and celestial cues; recognition of a home and of

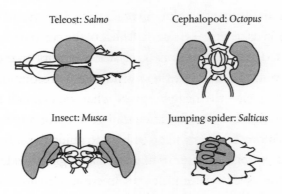

Fig. 11. The relative sizes of the visual regions of the brain (shaded) in four highly visual animals: a fish, an octopus, a fly, and a jumping spider.

food sources; and the detection of prey, predators, and potential mates.[11] This can be extended in particularly advanced groups to include recognition of individuals and the manipulation of tools; even, in humans, to sports and entertainment. The crucial point here is that the same image information has to be extracted in ways that allow it to be processed differently for varying purposes. This requires a great deal of brainpower.

Many advances in eye structure were required to complete the transition from Stage 3 to Stage 4—all the 'inimitable contrivances' mentioned by Darwin. However, one was critical, and it is the only one I will discuss in this chapter. For camera-type eyes, this was the evolution of a lens that produced a decent image, and that was not a simple optical task. Nearly all marine creatures with single-chambered eyes have spherical lenses, as these have a shorter focal length and higher light-gathering power than any other shape. But a spherical lens made of a single homogeneous material, such as glass or protein, produces an image so poor as to be unusable (Fig. 12a). The problem is known as spherical aberration. In a well-behaved camera lens, all rays from a distant point come together at the focus to give a sharp image of that point, and this is achieved by using multiple surfaces and different kinds of glass. With a simple spherical lens, a sharp image is not produced: the rays that have come through the centre of the lens are focussed at one point, but those from the outer regions are focussed much closer to the lens, and this mismatch gets worse the farther out the rays are from the centre. The result is that there is no focus, but a blur in which nothing is resolved. This problem is known as spherical aberration. The answer, which has been 'discovered' in all the four phyla that use this kind of lens, is to produce a lens which is not optically homogeneous: one in which the refractive index, which controls the amount by which the light is bent, decreases from the centre to the periphery. In such a lens, the outer rays are bent less strongly than the central rays, and in this way, all the rays can be brought to a single focus (Fig. 12b).

The origins of the idea of an inhomogeneous lens go back to Thomas Young, a remarkable nineteenth-century polymath, who, among other things, developed the three-colour theory of colour vision and demonstrated the wave nature of light. In a Royal Society lecture, in 1801, he proposed that

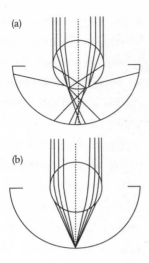

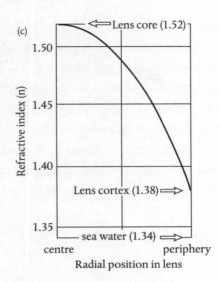

Fig. 12. Making a lens that produces a sharp image. a) Ray paths through a lens made of homogeneous glass or protein. b) Ray paths through an 'aplanatic' lens in which the refractive index decreases from centre to periphery. c) The ideal refractive index gradient in a lens capable of producing the ray paths in b).

the human lens is denser in the centre, and he made some calculations about how this would affect its focal length, but at that time, he did not have decent measurements of the actual refractive index gradient involved. Half a century later, James Clark Maxwell, apparently 'while contemplating his breakfast herring' in 1854, concluded that fish lenses have a refractive index gradient which eliminates spherical aberration. He went on, from this, to devise a 'fish eye universe' in which all rays starting from one point pass through a second point—the universe acting as a perfect lens. (Unfortunately this required implausible refractive indices.) In the 1880s, Ludwig Matthiessen measured the focal lengths of the lenses of fish and cephalopods, showed that these were inconsistent with homogeneous material, and produced a model with a parabolic gradient of refractive index which he believed would do the job. This was close, but not quite right, and it wasn't until 1944 that Rudolf Luneberg came up with the correct gradient, shown in Figure 12c.

This inhomogeneous, or graded-index, lens design is almost universal in marine animals with single-chambered eyes. It seems that once an eye starts to evolve in this direction, it rapidly reaches the right solution. When one considers the materials involved—basically more-or-less hydrated protein—this seems extraordinary. How is the appropriate gradient of protein dilution, responsible for the refractive index differences, set up? Most remarkably, how is the gradient across the lens maintained as the lens enlarges? In a fish the lens may increase in size by a factor of ten or more as the fish grows, and this means that at every point in the lens the refractive index must change continuously if the gradient *relative to the radius* is to stay the same. So far there are no answers to these questions.

2

COMPOUND EYES AND INSECT VISION

The large numbers of tiny lenses that cover the surface of a compound eye are not visible to the naked human eye. It was only with the advent of microscopy in the seventeenth century that the external structure of compound eyes could be seen and studied. Finding out how they worked took much longer, and it wasn't until the 1890s that this was achieved, and even then, there were gaps in the story that took another hundred years to resolve fully. One of the earliest scientists to depict a compound eye was Robert Hooke, whose book *Micrographia*, of 1665, contains a quite stunning picture of the eye of the 'Grey drone-fly', which modern writers identify as a male horse-fly (Tabanidae; Fig. 13). The eye consists of large numbers of hemispherical 'pearls' (Hooke's word for convex facets) arranged in a triangular lattice. As is typical of many male flies, the two eyes touch each other at the midline, and they are divided into two parts, with much larger facets in the upper half. In females the facets are of similar size across the eye, similar to the smaller facets of the males. The function of the large facets in the male is to provide higher resolution for spotting females in flight. It is the females that bite; the males feed on nectar—and chase females. Hooke not only drew the eyes but went on to provide a convincing theory of how they worked. Essentially this is the same as the 'mosaic' theory that Johannes Müller published a century-and-a-half later in 1826. Müller often gets the credit for this, but that actually belongs to Hooke.

Hooke was a remarkable polymath. Apart from the *Micrographia*, which is still in print, he made a portable watch, inventing the balance spring and a new type of escapement. He established Hooke's Law, which relates length and tension in elastic materials. He was a founder member of the Royal Society and from 1662, its Curator of Experiments. He was Surveyor of the

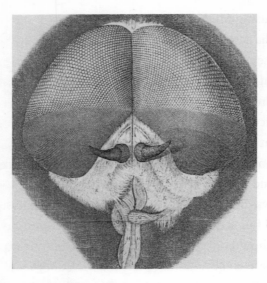

Fig. 13. The Grey Drone Fly. From Robert Hooke (1665). *Micrographia*. See also Plate 2.

City of London after the great fire of 1666, and collaborated with Christopher Wren in the reconstruction of public buildings, including St Pauls. He quarrelled with Newton over the discovery of gravity, to which Hooke seems to have the prior claim, but with Newton as president of the Royal Society, this did his reputation no good. No authenticated picture of Hooke survives (there had been one; its loss could have been Newton's doing), but we have this wonderful pen sketch of him by John Aubrey,[1] the antiquary and biographer, who was Hooke's friend and drinking companion.

> He is but of middling stature, something crooked, pale faced, and his face but little below, but his head is lardge, his eie full and popping, and not quick; a grey eie. He haz a delicate head of haire, browne, and of an excellent moist curle. He is and ever was temperate and modest in dyet etc.

The sketches of the apposition compound eye in Figure 8e (Chapter 1) and Figure 15 show that image formation is in two parts. Each facet-lens produces its own inverted image of a very small part of the outside world (Fig. 14), but the whole image of the world that the insect sees is upright and is not optical, but is formed, using Müller's word, by the *mosaic* of the contributions of all the ommatidia across the eye. An ommatidium is the unit consisting of a lens and the receptors beneath it. There was much confusion, really until the 1890s, about the relative roles of the individual facet images and the overall mosaic image. A famous cartoon by Gary Larson,[2] 'The last thing a fly ever sees', nicely illustrates the problem. It shows a hexagonal array of identical upright pictures of a woman with a fly swatter. Whilst the joke is good, the cartoon manages to incorporate most popular misconceptions about compound eyes. (Larson is actually a good biologist, and knew exactly was he was doing.) In fact the images should be inverted, they should not be identical but slightly displaced from each other, and only the very centre of each should be represented, and then the intensity in that central region should be averaged. More of this shortly.

The inverted images produced by each lens really do exist. They were first seen by Antonie van Leeuwenhoek, a Dutch trader from Delft who made microscopes using very small, powerful lenses. In a letter to the Royal Society of London in 1694,[3] Leeuwenhoek wrote:

Last summer I looked at an insect's cornea through my microscope. The cornea was mounted at some larger distance from the objective of the microscope as it was usually done when observing small objects. Then I moved the burning flame of a candle up and down at such a distance from the cornea that the candle shed its light through it. What I observed by looking into the microscope were the inverted images of the burning flame: not

Fig. 14. Recreation of Leeuwenhoek's observations. Images of a single candle flame, seen through the cornea of the eye of a robberfly (Asilidae). The outer face of the cornea is in air, and the rear face in saline, as in life.

one image: but some hundred images. As small as they were, I could see them all moving.

Figure 14 shows the images produced by an insect eye's multiple optics. The question that has to be resolved is how these many inverted images fit into the overall image, which, from the geometry of the eye, must be upright (Fig. 15). The images Leeuwenhoek saw are quite large, and if each were fully resolved by a retina like ours, they would present a mass of largely overlapping representations of the outside world. A way out of this confusion would be for the images to be curtailed, so that each is restricted to an angle of a few degrees, with each ommatidium making a different and unique contribution to the overall image. Although Hooke had not seen the individual images, he knew they must be there, and realized that the overlap problem needed a solution:

It follows, therefore, that onely the Picture of those parts of external objects that lie in, or neer, the Axis of each Hemisphere, are discernably painted or made on the Retina of each Hemisphere, and that therefore each of them can distinctly sensate or see onely those parts which are very nearly perpendicularly oppos'd to it, or lie in or neer to its optick axis. . . .The representation of any object that is made in any other Pearl, but that which is directly, or very nearly directly, oppos'd, being altogether confus'd and unable to produce a distinct vision. (*Micrographia: Observ. XXXIX*)

Hooke is saying that a single facet (hemisphere, pearl) and its receptors only see a small part of the surroundings, and that neighbouring facets do not see the same part, but (by inference) adjacent parts. He was certainly right about this. However, in his account, he attributes this directional restriction mainly to the fact that the resolution of a lens gets worse away from the optic axis. This does indeed happen, but the 'good' region of the image is too large for the field of view of a single facet-lens to be restricted by this. Müller, in 1826, attributed the directional specificity of each ommatidium to the shading effect of the pigment that lines the tube between the corneal surface and the retinal element, rather as in the simple compound eye in Figure 8 (Chapter 1). This is certainly part of the answer, but it leaves no real role for the facet lenses, nor the images they produce. Unfortunately, Müller did not know of Leeuwenhoek's images, not indeed of Hooke's work. In the 40 years after Müller's book, the tiny images were repeatedly rediscovered, and this almost led to the abandonment of the mosaic idea, even though there was no sensible replacement for it.

The man who revived the mosaic idea and made it secure was Sigmund Exner, whose 1891 book *The Physiology of the Compound Eyes of Insects and Crustaceans* is not only highly original but has become the basis for all later compound eye research.[4] Like his teacher, Hermann von Helmholtz, Exner was a multitalented scientist who is perhaps better known as a psychologist, publishing on both the location of function in the brain and the psychophysics of colour vision and apparent motion. In some ways his compound eye research was an interlude in his psychological career. He had three brothers, two of whom were physicists, and one of them, Karl, helped Sigmund with his optical ideas. His sister married Anton von Frisch, and their son, Karl von Frisch, went on to win a Nobel Prize for his work on the dance language of bees. In 1917, in recognition of Exner's distinction in his multiple fields, the Austrian government made him a knight, with the splendid title Siegmund Ritter Exner von Ewarten.

Exner's compound eye studies built on much improved histological studies in the 1860s and 70s, which finally revealed the real nature of the sensory elements in the insect retina. The crucial finding, for Exner, was that the receptive structure beneath each facet was a 'fused rhabdom', which we now know to be made up of microvilli contributed by eight receptor cells arranged around it. This rhabdom (from the Greek for 'rod') acts as a light-guide, which means that it accepts light only from the centre of the image provided by the facet-lens, and then more or less scrambles it (Fig. 15). So each ommatidium accepts light over a small angular cone and averages it, preventing any further resolution of each Leeuwenhoek image. In Exner's scheme the real functions of each lens are not to produce images but to define the field of view of each ommatidium—its acceptance angle—and to greatly increase the amount of light that reaches each rhabdom.

The majority of insect eyes, including those of cockroaches, grasshoppers, bees, and dragonflies, are of this type and are referred to as *apposition eyes*—which was the name Exner gave to them—because the contributions of individual ommatidia to the overall image are apposed, or adjacent to each other. There are two other types of compound eye which we will meet later in the chapter. These are neural superposition eyes of two-winged flies, which are apposition eyes in which there is some resolution of each facet image, and optical superposition eyes, which have a fundamentally different optical structure and a single, erect, deep-lying image.

Apposition eyes are enormously successful, thanks largely to insect fecundity, and they have many desirable features. They can be spread out over the head surface, rather than requiring a deep pit, as our eyes do, and so they can be part of the body armour around the brain, rather than a vulnerable liability. But they do have one serious drawback compared to camera-type eyes like ours, which is that their resolution is very limited.

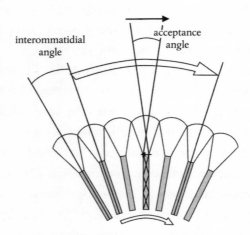

Fig. 15. The geometry of an apposition eye, as described in Exner's synthesis of 1891. Each facet lens forms an inverted image (small solid arrow, see Fig. 14), but only the centre of each image is accepted by the photosensitive rhabdom, where the light is trapped and scrambled by total internal reflection. In this way each ommatidium makes a contribution to the overall erect image (open arrows) that does not overlap with its neighbours. In general the acceptance angles of single ommatidia are similar to the inter-ommatidial angles between them.

A bee's eye resolves about 1°, and this compares very unfavourably with our own resolution, which is a hundred times better—less than one *minute* of arc. This difference in resolution is a consequence of the very small size of the lenses that make up a compound eye. The phenomenon involved is known as diffraction. When light from a distant point is brought to a focus by a lens, it doesn't produce a point image, as one might expect, but rather a small blurred disc, and, counterintuitively, the smaller the lens, the wider this disc becomes. The reason for this is that on a small-scale light behaves as a wave. When focussed light reaches an image, the waves from different parts of a lens interfere with each other—some adding and some cancelling—and the result is a circular light pattern known, after its discoverer, as an Airy disc. The smaller the lens, compared with the wavelength of light, the wider the disc, and conversely, the bigger the lens, the smaller the disc—hence the enormous telescopes used in astronomy. Fortunately, there is a simple formula for working out the width of this disc. The angle the disc makes, at half its maximum intensity, when projected onto the world outside the eye, is given simply by λ/D (in radians: 1 radian is 57.3°). Here λ is the wavelength of light (0.5 micrometres or μm for green light, where 1 μm is a millionth of a metre) and D is the facet diameter. For a bee, with 25 μm diameter facets this comes to 1.15°. In our eyes, in daylight when the pupil is 2.5 mm wide, the Airy disc is one hundred times smaller, about 0.7 *minutes* of arc. To get an idea of how the world would look to a bee, imagine the scene divided into pixels about the size of the nail of your index finger at arm's length: all the fine detail of leaves, twigs, and letters will be lost, but not the coarser features. This deficiency does not stop a bee from finding food, navigating by landmarks, or coordinating its flight. Despite poor resolution, eyes can still be impressively useful.

How might a bee's resolution be improved? This is where the compound eye's real design flaws become apparent. Making smaller facets, to fit more in, clearly won't work because the resolution of each image will get worse, not better. To double the resolution, reducing the width of each Airy disc to, say, 0.6°, will require the doubling the size of each facet to 50 μm. But this will now mean that, to exploit the improved resolution, *twice as many* ommatidia

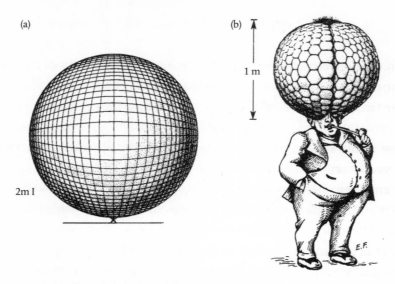

Fig. 16. a) Mallock's and b) Kirschfeld's versions of a compound eye with human resolution.

will be needed *and* their lenses will be twice as big. From this it follows that the size of a compound eye increases as the *square* of the resolution it produces, whereas in a camera-type eye, the single lens only needs to increase in size in simple proportion to the resolution. The first person to notice this limit to compound eye resolution was Henry Mallock, an optical instrument maker who understood the limitations imposed by wave optics. In his paper of 1894, he worked out the size a compound eye would need to be if it were to have the same resolution as a human eye (1 arc minute), and he concluded that it would have a diameter of 28 *metres* (Fig. 16). In 1976 Kuno Kirschfeld reworked Mallock's calculation,[5] pointing out that the assumption of high resolution everywhere is a little unfair, because the human eye only has 1-minute resolution in the fovea, and for most of the retina, the resolution is less than a tenth of this. His redrawing of Mallock's man/insect now has an eye 1 metre across (Fig. 16), but even in its modified form, it is still not something one would want to fly around with.

Special Zones in Insect Eyes

The resolution restriction imposed by diffraction is a real problem for insects, and many try to squeeze in at least a small region of higher resolution—and larger facets. There are two reasons for needing enhanced resolution: sex and predation. In many two-winged flies, such as the horse-fly shown in Figure 13 and Plate 2, and also in drone bees, the males have a region of enlarged facets pointing forwards and slightly upwards which they use to spot and then chase females, which they typically mate on the wing. Tom Collett and I studied a particularly nice example of this behaviour in the 1970s.[6] The animals were small hoverflies, *Syritta pipiens*, which fly around flowers in gardens in summer, and can be persuaded to behave fairly normally in a large transparent box, where they can be filmed. The males and females are of similar size, but the males have eyes with a forwards-looking region of much larger facets. In this region the resolution (the angle between ommatidia) is 0.6°, rising to 1.5° in the rest of the eye; in the females the resolution is 1.5° across the whole eye (Fig. 17). Females fly around in a jerky manner, with straight segments of flight punctuated by quick body turns. When a male detects a female, he shadows her smoothly at a distance of about 10 cm, keeping her directly ahead (Fig. 18). The beauty of the shadowing strategy is that the male can actually remain out of sight of the female, because his resolution is much better than hers. The tracks of the two flies show features that are interestingly similar to the ways that humans move their eyes, although here it is the whole body that moves rather than the eye in its socket. Unlike most animals, and indeed most insects, hoverflies do not rotate their heads on their bodies during flight, so that the direction of the axis of vision is controlled entirely by the wings acting on the body: forwards eye direction and body direction are the same. The track of the female in Figure 18 shows long sequences in which the body moves forwards, or sideways or some combination, but without rotation. Then, at irregular intervals, the body makes an abrupt turn, for example, halfway between 1 and 2, at three frames before 3, then nothing until 5, and so on.

The angular direction of vision is either stationary or rotating very fast. This is exactly what our eyes do for most of the time: they either *fixate* steadily or change direction by making fast movements known as *saccades*. The main function of this nearly universal 'saccade and fixate' strategy is to ensure that the eye stays still enough to avoid motion blur, which is what occurs when we take a photograph with a moving camera. But when the eye does have to move, it does so very fast, to keep the periods of blurred vision as short as possible. How this strategy is achieved depends on what mechanisms are available: for us, it is the eye muscles, for a hoverfly, it is the wings.

The track of the male is quite different. There are no abrupt turns, nor periods without rotation. The rotation is quite smooth and driven by the position of the female on the male's retina. It is reminiscent of smooth pursuit in humans, when the eyes and head rotate continuously to track some object; for example, a football, which does not move too fast relative to the eye. Interestingly, when a male *Syritta* is not tracking a female, his behaviour is just like that of the female. This is the same for us: without something to track, we revert to saccade and fixation behaviour. The other extraordinary feature of the male's behaviour is the way he maintains a distance of close to 10 cm throughout the pursuit. This involves backing off at times and coming forwards again at others. How he does this is not clear. He has no binocular vision: the eyes are much too close to each other (Fig. 17), and

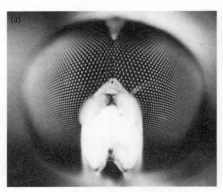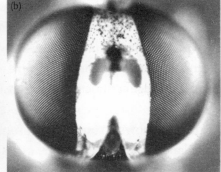

Fig. 17. a) The eyes of male and b) female hoverfly *Syritta pipiens*, showing the forwards-pointing region of enlarged facets in the male.

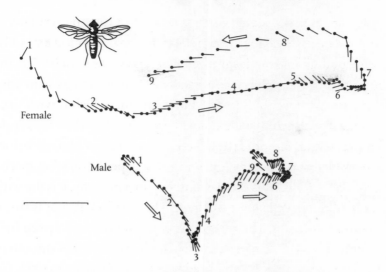

Fig. 18. Frames from a ciné film of a male *Syritta* (below) shadowing a female (above). Scalebar is 5 cm. The flies are viewed from above. Only the head and body axis are shown. The female has long stretches of flight without rotation, separated by abrupt saccade-like turns. The tracking male rotates smoothly and continuously, always keeping the female within 5° of his body axis. As the female approaches, he backs off, maintaining a distance of about 10 cm, before coming forwards again as she passes and turns away. The frame interval is 20 ms, with corresponding points on the two tracks numbered. The insert is an enlarged drawing of a female *Syritta*.

their fields barely overlap. The only reliable clue seems to be the size of the female's head; the body is no use because its apparent length changes each time the animal turns. The head has a diameter of about 2 mm, which at 10 cm makes it just over 1° across at the pursuer's eye, and this means that it is seen by about four ommatidia. This is not much, but probably enough for the male to judge contraction or expansion, and move forwards or backwards accordingly.

Figure 19 shows an amusing example of what happens when the female finally lands. The male hovers briefly, then accelerates rapidly, and more or less collides with her on the flower or leaf. The mating dash in Figure 19 has an interesting twist: a second male on the right who had been shadowing the first male continues to track him up to his contact with the female. It is

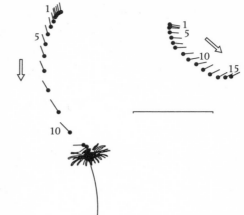

Fig. 19. Frames from a film of a male *Syritta* darting onto a stationary female on a flower, watched by a second male (right). Scalebar is 10 cm. Filmed from above. The frame interval is 20 ms.

interesting that at frame 10, the observing male's direction is about two frames (40 ms) behind the moving male, and this presumably reflects the delay in his tracking mechanism. This behaviour is probably not overt voyeurism: it seems that males cannot tell the sex of another fly on the wing, and confusion is common.

Some of the larger hoverflies, such as *Eristalis* and *Volucella*, have a quite different strategy. They hover under trees, often in sunbeams, and will attempt to intercept anything that comes past. Nine times out of ten, these interception courses are inappropriate, and the chase is broken off early. However, if the object being intercepted is the right size and moving at the right speed, there is a chance that it is a female and can be caught after a short chase. Tom Collett and I wanted to know how the flies calculated their interception course, given that the potential object of interest might appear anywhere in the space around them. This study was greatly helped by the tendency of the hoverflies to return to the same aerial station after each pursuit, presumably by matching visual landmarks. This meant that one of us could lie on the ground below this patch of space, with a camera pointing upwards. Meanwhile the other fired projectiles of various kinds and sizes, from peppercorns to small potatoes, in the direction of the hoverfly, using a variety of peashooters and catapults. We also used tethered peas on

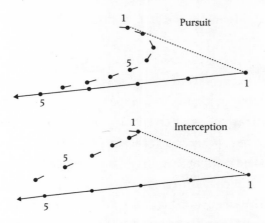

Fig. 20. A pursuit flight and an interception flight. In pursuit the fly heads constantly towards the target, with a slight delay. During interception the initial turn is away from the target, and after this, the track is straight. The interception course takes less time to reach the target (see the relative positions of frame 5). The dotted line shows the initial sighting point.

strings, which stopped after pursuit had started, so that we could see how pursuit broke off. Sadly, I have no photographs of us involved in these weird procedures. The outcome of the study was that these were true interception flights, with the fly making a turn on the spot and flying off at top speed in a direction calculated from the position at which the 'target' first appeared, and its apparent velocity. Unlike ordinary chasing, where the pursuer turns so as to keep the target ahead (as with *Syritta*), during interception the initial course of the pursuer is often away from the actual position of the target (Fig. 20). With such a strategy, the pursuer can get to a point on the target's future track, rather than to where it is now. We came up with a formula that predicted the fly's initial course, whatever the nature of the projectile, but crucially this course would only lead to interception if the projectile was first seen at a distance of about a metre and was travelling at about 8 metres per second.[7] These figures are consistent with the size and speed of a female. Larger targets were seen too early, and smaller targets too late, and in neither case did the interception work. What this shows is that the behaviour of the males acts as a kind of filter, so that the conditions for effective interception are only met if the assumptions built into the strategy fit the intended female quarry.

In some other insects, notably robber flies and dragonflies, there are regions of larger facets and enhanced resolution in both sexes. In these the reason is not sex but predation. The larger (hawker) dragonflies have a narrow band of high resolution directed upwards, and as they patrol over fields and lakes, they detect other insects against the sky (see Fig. 24). They then swoop upwards and capture their prey from below. Other dragonflies perch on branches and launch themselves upwards to intercept their prey. A recent study by Matteo Mischiati and his colleagues on a perching dragonfly found that they, like hoverflies, use an interception strategy, but because dragonflies have heavier bodies, the way they go about it is different. The dragonfly launches in the direction of the prey, but while the body takes some time to come onto a course in line with the prey, the animal's head turns first and maintains fixation on the prey with its high-resolution zone. The fixation of the prey by the head is impressively fast and accurate, implying that this is more than a matter of simple reactive reflexes. It seems that, for the head to track with such precision, the dragonfly must use a mechanism involving prediction, both of its own body movements and of the prey's future track. This is similar to the way that humans, when tracking a moving target, quickly switch from a rather slow reflex mode to one that predicts future target motion.

Let's now return to the optics of fly eyes, which, until the 1960s presented something of a mystery. The two-winged flies (Diptera), examples of which are houseflies, blowflies, robber flies, and hoverflies like *Syritta*, have an ommatidial arrangement subtly different from other apposition eyes. It is one in which the inverted images produced by the facet lenses (Fig. 14) are actually resolved by a mini-retina—raising again all the problems that Exner's synthesis (Fig. 15) had tried to resolve. What happens in dipteran eyes is that the individual photoreceptor elements do not fuse together to form a single rhabdom, as in a conventional apposition eye, but stay separate as individual *rhabdomeres*. This means that the seven individual rhabdomeres receive different parts of the image that the facet-lens provides. Does this mean that the fly in some sense 'sees' both kinds of image—the inverted and the erect images in Figure 15? The answer, fortunately, is no, but the reason for this arrangement is equally surprising. The pattern of rhabdomere endings was first illustrated with exemplary accuracy by Hermann Grenacher in 1879, in his book on the eyes of arthropods. His engravings are masterpieces of clarity, and I am always astonished at the accuracy with which he drew things whose function was not revealed until many decades later. His tiny illustration of a cross-section of two ommatidia of a horse-fly (Hooke's insect) is no exception (Fig. 21b). Modern electron micrographs of fly receptors show the same pattern of six rhabdomeres arranged in a slightly odd triangular arrangement, with a seventh projecting into the central space. (In fact the central rhabdomere is a pair, one above the other.) One feels that a lesser anatomist than Grenacher would have made the pattern look hexagonal—because nearly everything else is—but he didn't: he drew its peculiar geometry exactly as he saw it. The reasons for the odd geometry are still obscure, but the reasons for having separated rhabdomeres which resolve the image are now understood, although the explanation is not what one might expect. Our current understanding is thanks mainly to the studies of Kuno Kirschfeld and Nicolas Franceschini in the late 1960s and early 1970s. They found that the axis of each rhabdomere in one ommatidium is precisely aligned with the axis of one of the other rhabdomeres in an adjacent ommatidia (Fig. 21a). In other words, seven rhabdomeres in *separate* ommatidia share the same field of view. Then comes the strange part. The neurons from all the receptors whose rhabdomeres image the same direction in space meet up in a single element in the next neural level—the lamina. This involves an impressive piece of neural engineering: a rotation and separation of the bundles of neurons from the receptors, and a precise regrouping (Fig. 21c). The result is that the neurons of the lamina behave as though they were components of an ordinary apposition eye, responding simply to light from the different directions A, B, C, etc., as though none of the complicated overlapping connexions between ommatidia had happened (Fig. 21a). On the face of it, all

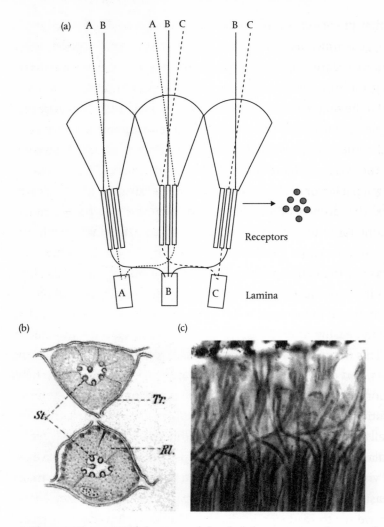

Fig. 21. a) The arrangement of the axes of the rhabdomeres in a dipteran fly. The direction of view (B) of the central rhabdomere of the middle ommatidium is the same as that of the outer rhabdomeres in adjacent ommatidia. Similarly for (A) and (C). The inset on the right (arrowed) shows the actual rhabdomere pattern in one ommatidium. The neurons from receptors with the same field of view all meet up in the next neural layer, the lamina. b) Grenacher's engraving of 1879 of the rhabdomere pattern (St) in two ommatidia of a horse fly. c) The twist in the receptor bundles and reorganization of receptor neurons between the retina and lamina.

this high-precision complexity appears to have achieved nothing; the over-all image is much as it would have been if each ommatidium had a single rhabdom. Clearly, this can't be the whole story. The answer seems to be that having multiple receptors imaging the same direction allows a stronger signal to be sent to the brain via the lamina, effectively equivalent to having a single receptor that obtains more light. And because this augmented signal comes from several rhabdomeres all with the same, narrow field of view, there is no reduction in the eye's resolution. In fact, the gain in signal strength is a factor of six rather than seven because the neurons of the small, central rhabdomeres (inset, Fig. 21) go straight through the lamina and do not contribute to the signal summing; they represent a high-resolution colour system a little like the cone system of humans. Consider how else an increase in the light signal might be produced. An equivalent increase in signal in an ordinary apposition eye (Fig. 15) could be achieved by getting each rhabdom to increase its light absorption by a factor of six. That would mean increasing its cross-sectional area by 6, and so its diameter by $\sqrt{6}$, or 2.45. This in turn would increase the acceptance angle of the ommatidium by 2.45, and decrease the eye's resolution by the same factor. The dipteran solution gets round this, increasing the light signal without compromising resolution.

Is all this worth the effort? In dim light, in the morning and evening or in woodland, the low number of photons that a receptor receives starts to limit vision. Small numbers of photons, arriving at random, make for an increasingly 'noisy' signal. We are all aware of the consequences of this when we try to read in poor light. For an insect the same is true. I asked Kuno Kirschfeld what he thought the six-fold gain in signal strength might mean for a fly, and he said it meant about a quarter of an hour at dawn and dusk. When I said that this didn't seem much he pointed out that this is when many flies swarm to mate, and it is the time when the eyesight of predators—mainly birds—is becoming poor because their eyesight, too, is limited by low photon numbers. In survival terms, this relatively safe win-dow is probably very important.

The Pseudopupil

An intriguing phenomenon, seen in the eyes of many insects and crust-aceans with apposition eyes, is a black dot which appears to move across the eye surface when you view the eye from different directions (Fig. 22 left). Sometimes this phenomenon, called the pseudopupil, is more elaborate, with a hexagonal pattern of spots and other markings (Fig. 22 right). The movement of the pseudopupil, apparently evoked by your own movement, gives the impression that the eye is watching you in some sinister way. The explanation is actually more prosaic. The ommatidium that shares a line of sight with your eye must appear dark because it is absorbing light from your direction, whilst other ommatidia do not, and remain light. As you move round the eye, different ommatidia come to share your line of sight, and so the black dot appears to move, but there is no movement other than your own. There is a little more to it, as Figure 23 explains. The pseudopupil you see most clearly is not at the eye surface, but deeper within the eye, and it is formed by the superposition of the images of all the structures that are at the focus of the individual facet lenses. At the centre of this will be the dark central rhabdom, but there is often also a hexagonal pattern of pigment cells of different colours. In butterflies this pattern can be particularly intri-cate and beautiful (Fig. 22 right and Fig. 23). In many cases the focal plane of each lens extends not just to its own rhabdom and pigment pattern, but also to that of neighbouring ommatidia, and in that case, multiple pseudo-pupils arise in a hexagonal pattern that can extend over much of the eye, as in the butterfly example. If you find this all rather difficult, you are in good company. Exner himself remarked: 'I come to the discussion of a remark-able optical phenomenon, the unravelling of which has cost me much puzzlement.'

Besides being a curiosity, pseudopupils can be very valuable to a physi-ologist. A line joining the pseudopupil to the observer's eye—or microscope—defines the line of sight of the ommatidium at the pseudopupil centre. By rotating the eye through a known angle, and seeing by how many omma-tidia the pseudopupil moves, one can work out the inter-ommatidial angle

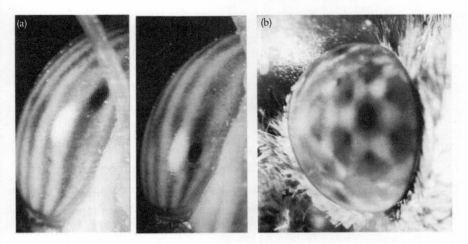

Fig. 22. Pseudopupils. a) (two panels) Simple pseudopupils in the eye of a locust, seen from two directions. b) Complex multiple pseudopupils in the eye of a butterfly *Junonia villida*.

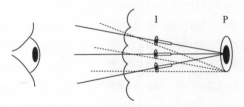

Fig. 23. Explanation of the deep pseudopupil. An observer (left) sees the combined and magnified image at P of all the structures in the focal planes (I) of a number of facet lenses. This combined image is deep in the eye, near its centre of curvature.

(Fig. 15) and so the local resolution of the eye. If the pseudopupil moves across five ommatidia when the observer moves 10° around the eye, then the inter-ommatidial angle must be 2°. In this way one can build up a map of the pattern of resolution across the eye, and of the way it differs between horizontal and vertical. This requires a fairly elaborate contraption for rotating either the microscope or the eye. Figure 24 shows some maps produced in this way.

The maps represent a globe around the animal's head, and show the eye's resolution in different directions, with resolution being measured as the number of ommatidial axes covering a given area of space. The first is

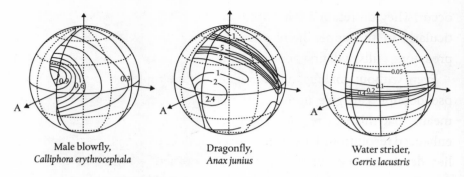

Male blowfly,	Dragonfly,	Water strider,
Calliphora erythrocephala	*Anax junius*	*Gerris lacustris*

Fig. 24. Maps of the distribution of resolution mapped onto a globe around the left eyes of three insects (*Calliphora erythrocephala*, *Anax junius*, and *Gerris lacustris*). The contours represent numbers of ommatidial axes per square degree. The different patterns are discussed in the text. A is the anterior (forwards) direction.

of the (left) eye of a male blowfly, and it shows the forwards-pointing high-resolution region the male uses when chasing females—Roger Hardie calls this, rather poetically, the 'love spot'—and it is typical of dipteran flies that go in for aerial mating chases. The facets in this region are usually larger than elsewhere, as in the hoverfly *Syritta* (Fig. 17) and the male horse-fly in Plate 2. The right-eye resolution map will be the mirror image of the left, so that the left and right maps overlap across the central midline, enhancing vision in this crucial region. The second map is of the eye of a large dragonfly, and it shows a long but quite narrow band of very high resolution across the whole of the upper hemisphere, imaging a strip of sky above and ahead of the animal. One can imagine the dragonfly using this rather like the line on an old radar set, to detect potential insect prey against the bright background of the sky before swooping upwards to make a capture. The third map is of *Gerris*, the pond skater or water strider. This is an animal which lives and hunts on water surfaces. Its resolution distribution reflects this, and it is almost confined to a narrow horizontal band covering a few degrees above and below the two-dimensional surface of the pond. Eyes of other 'flatland' animals are similar. Ghost crabs (*Ocypode*) and fiddler crabs (*Uca*), both of which live on sand or mud flats, have greatly enhanced resolution in a region around the horizon. This is where their social interactions with mates and rivals occur. They do retain a reasonable degree of resolution elsewhere, particularly in the upper hemisphere, because this is where their main predators, gulls, will first appear.

Even without making measurements, the simple appearance of the pseudopupil can give useful information. A locally large pseudopupil means that this is a region of smaller inter-ommatidial angles, and so enhanced resolution. This is because the visible pseudopupil (P in Fig. 23) lies deeper in the eye and is more magnified. In flatland crabs the pseudopupil around the eye's equator is very tall and thin, which means that the vertical resolution is much higher than the horizontal. In an eye of limited dimensions, this is a compromise that enhances detail in objects that matter, although it must give the crabs a peculiarly astigmatic view of the world.

Superposition Eyes of Beetles

In his 1891 book,[4] Sigmund Exner describes a novel and remarkable optical system in the eye of the European glow-worm (*Lampyris*). Male glow-worms fly around at night in the summer months, looking for the green glow from the abdomens of females. The females never become adult winged insects—unlike their relatives, the fireflies. Unlike the males they are almost eyeless and crawl around in low vegetation, advertising their presence with a steady light. If you clean the retina and other debris from the inside of the eye of a male glow-worm or firefly, leaving only the optical array and then suspend this from a drop of fluid with air on the outside (as in life), you can use this eye surface as a lens. With it you can view images of objects in the world, and take photographs of them, as Exner himself did. The most astonishing thing about these images is that they are *erect*—the right way up (Fig. 25). Camera lenses, and all other lenses, produce images that are *inverted*, as are the images produced by the facet lenses of normal apposition eyes (Fig. 14). Exner realized that there was a difficult optical problem here, and set about solving it.

Fig. 25. Photograph of the image of a famous nineteenth-century naturalist in the superposition eye of a firefly (*Photurus* sp.). Unlike the inverted multiple images in apposition eyes (Fig. 14), in the firefly, there is a single erect image, as Exner first demonstrated.

Exner's own section of a glow-worm eye is reproduced in Figure 26. The key features are the outer region consisting of individual optical elements, a wide zone with little material in it across which rays can converge, and a retina lying beneath this clear zone, a long way from the optics and roughly half way out from the centre of the eye. This is quite unlike an apposition eye, where the receptive rhabdoms make contact with the rear of the optical structures (Fig. 15). It is easy to show that an array of simple lenses will not produce any sort of image on a deep-lying retina. Exner concluded that such an image could be formed if the optical elements behaved not as lenses but as inverters: that is, devices which change the direction of a beam of light so that it emerges on the same side of the axis as the incoming beam. In this they differ from ordinary lenses where the beam emerges on the opposite side of the axis. The consequences of this arrangement are shown on the right of Figure 26. The inverted rays all converge to a point deep in the eye, *and the overall image* (AB) *is erect*. The next question is how to make an inverter. With ordinary glass lenses, this is straightforward. One lens brings

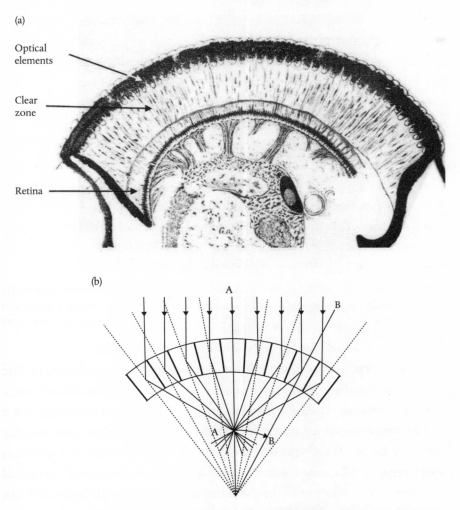

(a)

Optical
elements

Clear
zone

Retina

(b)

A

B

A

B

Fig. 26. a) Section of the eye of a glow-worm (from Exner), and b) a simplification of Exner's diagram of the ray paths needed to produce an erect image. The function of the optical elements is to invert the paths of the entering rays so that they emerge at the same angle, but on the opposite side of the element's axis. This arrangement forms an image of rays A and B from a distant object at the arrow AB on the retina.

a parallel beam of light to a focus. A second identical lens, placed at two focal lengths from the first, takes this focussed light and turns it back into a parallel beam (Fig. 27c). This effectively makes a telescope with no magnification. If the original beam is parallel to the axes of the two lenses, it will emerge from the second as an identical parallel beam: nothing seems to have been achieved. But if the beam makes an angle with the axis of the first lens then it will emerge from the second lens making the same angle, but in the opposite direction relative to the axis (Fig. 27c). The beam, though otherwise unchanged, has been inverted. In principle a glow-worm eye could consist of an array of such two-lens telescopes.

This is where Exner made his second extraordinary discovery. He realized that the curved surfaces in the optical elements of the glow-worm eye did not have enough ray-bending power to make the two lenses needed to produce an inverter. He proposed instead an optical structure that he called a lens-cylinder. This consists of a transparent cylinder made of circular layers of material whose refractive index changes from high in the central region along the axis, to lower values as the distance from the axis increases. The exact refractive index gradient is important: this decrease has to have a nearly parabolic profile, as Exner's physicist brother, Karl, worked out in 1886. What happens in such a cylinder is that rays that start off parallel to the axis are refracted continuously towards the centre, as they encounter a higher refractive index on one side and a lower index on the other. In this way a parallel beam of light is brought to a focus, and the structure behaves as a lens (Fig. 27a). As Exner showed, this is what happens in the corneal lenses in the apposition eye of the horseshoe crab *Limulus*. Imagine then a lens-cylinder twice as long as the distance to the first focus. The second half of the cylinder takes the focussed light, and turns it back into a parallel beam (Fig. 27b), exactly as in the two-lens telescope version (Fig. 27c). If the components in the glow-worm eye were to contain the right refractive index gradient, then the mystery of how to produce an array of inverters would be solved.

Exner tried very hard to demonstrate the presence of refractive index gradients in these tiny structures, but the technology available to him in the 1880s was not really up to the job. He had made a device (a micro-refractometer) to measure the refractive index of very small objects, and was able to show that there was indeed a radial gradient within these optical components: these structures in glow-worm and water beetle eyes were not

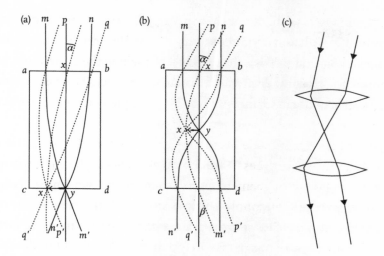

Fig. 27. a) and b): Exner's own diagrams of the paths of rays through single- and double-length lens cylinders. Rays are bent continuously within the structures. Single-length lens cylinders form inverted images in some apposition eyes, and double-length lens cylinders act as inverters in superposition eyes. c) shows a two-lens inverter, analogous to the lens cylinder in b).

simple lenses. Crucially, he showed that flat-sided sections of *Limulus* lenses, still behaved as lenses: with no curved surfaces, lens-cylinder optics was the only possible alternative. But he could not measure the form of the gradient within these structures. In fact, this was not achieved until 1973, more than 80 years after the publication of Exner's book. In the meantime, Exner's ideas had mixed fortunes. Many authors did not understand them (understandably!), and textbook accounts often reflect this confusion. During the 1960s doubts again arose about the plausibility of lens-cylinder optics (even though such devices had already been produced commercially), and other ideas about how 'clear-zone' eyes might work were rife. Many of them didn't make much sense. The situation was resolved in 1973 when Klaus Hausen, working in Tübingen, was able to measure the refractive index gradient in sections of the optical elements from the eye of a moth (*Ephestia*), using a relatively new device called an interference microscope. He found a parabolic gradient of refractive index within the optical elements, exactly as Exner had predicted. This impressive vindication of what had until then been essentially a brilliant conjecture seems to have brought to an end any further speculation about how superposition eyes work. There is one important exception to this. The long-bodied crustaceans (prawns, crayfish, and lobsters) have a quite different kind of superposition eye based on mirrors, which I shall explore in Chapter 3.

Moths and Butterflies

In choosing to work with a glow-worm eye, Exner had been astute. The optical components of glow-worm eyes are fused to the cornea and so don't come away when the cornea is cleaned up to make a lens. This is not generally true. In moths, the other great insect group with superposition eyes, the optical elements have two components: the cornea itself and beneath this, a long, glass-like, bullet-shaped structure known as the crystalline cone (shown in Fig. 29). These are only loosely attached to the cornea and come away from it easily. These optical structures can be studied in fixed or frozen sections but not together in an intact preparation, so it is not possible to make the sort of photograph seen, for example, in Figure 25. Nevertheless, the structure of moth eyes is almost the same as glow-worm eyes (Fig. 26), and many other optical observations indicate that moth eyes have a superposition optical system no different in principle from the eyes of glow-worms, fireflies, and some other beetles.

One of the striking features of moth eyes is the way they glow at night in the beam of a torch. This eye-glow arises from the presence of reflecting pigment around the base and sides of the receptors. When light enters the eye, it is focussed and then reflected back from this reflective material, and the eye glows just like a cat's eye in the headlamps of a car. If light enters the eye as a roughly parallel beam, then it will be focussed to one or a few cells on the retina, and when these reflect light the optics will convert this light back into a parallel beam, emerging from the same facets through which the original beam entered; imagine the optical diagram in Figure 26, but with the arrows reversed. The size of the patch of glow tells an observer the

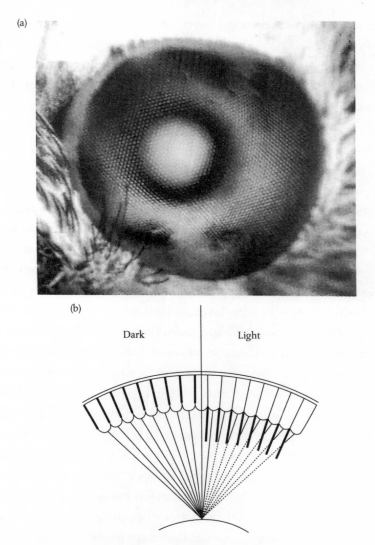

Fig. 28. a) Eye glow in a hummingbird hawk-moth (*Macroglossum*). b) Extinction of the glow as pigment migrates inwards from between the crystalline cones, cutting off all but the central rays.

size of the 'superposition pupil'—the amount of the eye surface that contributes rays to the image at any point on the retina. In the moth eye shown in Figure 28, the glow patch is about 16 facets wide, meaning that about 200 facets contribute light to the retinal image. This huge collecting area is what makes superposition eyes so useful at night, as it gives a much brighter image than an apposition eye can provide. This was presumably the reason why superposition eyes evolved from apposition eyes in the first place. Besides nocturnal insects, superposition eyes of this refracting type are found in many crustaceans that live in deep water, where light conditions are similar to terrestrial twilight. Good examples are the krill (euphausiids), marine shrimp-like crustaceans which swarm in great numbers in the southern oceans and are a major food source for filter-feeding whales.

The eye-glow of many moth eyes starts to disappear if the eye is illuminated for half a minute or more. This is caused by a movement of dark pigment inside the eye, migrating inwards from between the crystalline cones (Fig. 28b). This progressively cuts off the rays that cross the clear zone, causing the superposition pupil to darken and shrink to one or a few facets. This enormously decreases the amount of light reaching the retina, effectively reducing the eye to an apposition configuration, where only one facet illuminates one rhabdom; this protects the receptors from the brightness of daylight. In the dark the glow returns over many minutes as the pigment returns to its original position.

The Lepidoptera is one of the largest orders of insects. All lepidopterans are characterized by a covering of scales, and an extendable proboscis which they use to drink nectar from flowers. Of the 120 or more lepidopteran families, only two are butterflies: conspicuous, colourful diurnal insects whose main distinguishing characteristics are the swollen club-like ends of the antennae. In terms of eye structure, most moths are crepuscular or nocturnal and have superposition eyes, but butterflies do not. Each ommatidium is separated from the next by pigment, and the rhabdoms reach up to the base of the optical elements, as in the apposition eyes of most other diurnal insects (Fig. 15). There is a mystery here. The butterflies evolved from the main moth line in the Cretaceous Period (145–66 Mya), at the time of the radiation of flowering plants. Presumably the ancestors of butterflies had superposition optics. How, then, had the butterflies reverted to apposition optics, if this was indeed what had happened?

When I was in Canberra in the early 1980s, on a 2-year fellowship at the Australian National University, I met Dan-Eric Nilsson. Dan was interested in the butterfly eye problem, and we decided to work on it together. We were joined by Joe Howard, whose knowledge of physics turned out to be indispensable. The question really boiled down to the nature of the optical elements in the butterfly eye: were they just lenses, as in the eye of a bee or fly, or were they modified inverters, derived from their equivalents in moths? Canberra was a wonderful place for butterflies, and we settled on a large species, *Heteronympha merope*, which was everywhere in the summer months. After a while we accumulated various kinds of indirect evidence which suggested that the optical structures were not simple lenses, but we had no definite proof. We didn't have an interference microscope, which is what Klaus Hausen had used to show that moths had lens-cylinder optics, and the only real alternative was to show that flat-sided sections of the optical elements in butterfly eyes still behaved as lenses: the method Exner had used with *Limulus* eyes. The crystalline cones—the optically relevant structures—are very small, about 40 μm long and tapering to a width of about 5 μm (one two-hundredth of a millimetre). I recall Joe and I saying to Dan that these might be lens-cylinders, but that they were too small to be able to do anything with. Anyway, Dan disappeared for a couple of days, cutting sections and taking photographs. He emerged with the photograph shown in Figure 29. We were amazed. This 5 μm-thick slice from the bottom of the crystalline cone was a lens that produced an inverted image (re-inverted in the figure), so it must be acting as a lens-cylinder (as in Fig. 27a). From the magnification of the image, the focal length of the bottom 10 μm part of the crystalline cone was astonishingly short: 5 μm. In optometrists' terms this gives this lens a power of 200,000 dioptres: 10^5 times the power of reading glasses.

What was this enormously powerful lens-cylinder doing in what appeared to be an apposition eye? A full reconstruction of the optics showed that this was the lower part of a system that functioned as a two-lens inverter, like

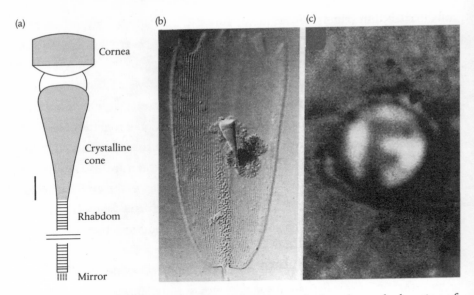

Fig. 29. a) Components of a butterfly ommatidium. The bar shows the location of the powerful lens-cylinder in the crystalline cone, and is 10 μm long. b) Crystalline cone on a moth scale, to indicate its size. c) Dan-Eric Nilsson's photograph of the image of a letter F, taken through a flat-ended 5-μm section from the bottom of a crystalline cone in the region indicated by the bar in a).

the inverters in moths but with the difference that these butterfly inverters had magnifying power (Fig. 30). The magnification increased the angle of rays leaving the optics by about 6.4, compared with the entering rays; this contrasts with a magnification of little more than 1 in the eyes of moths (Fig. 27b). This has two consequences. First, a ray entering a facet at an angle of 2° from the ommatidial axis will emerge from the tip of the crystalline cone at 12.8° (Fig. 30b). The significance of this is that 12.8° is about the maximum angle that a rhabdom can accept light. Because the rhabdom has a higher refractive index than the surrounding tissue, it will trap rays by total internal reflection, but only up to a certain angle: at any greater angle the rays will not be trapped, and lost to the screening pigment around the rhabdom. In this way the magnification of the optics and the refractive index of the rhabdom define the rhabdom's acceptance angle. This contrasts with an ordinary apposition eye, where the acceptance angle is set by the diameter of the rhabdom (Fig. 15). Second, the width of a parallel beam of light that enters a facet is reduced in diameter by the optical system to a width that just fits the rhabdom (Fig. 30a). These two effects mean that the butterfly eye has perfect apposition optics, but based on a quite different principle from that of a fly or bee. Because there is no focussed image of the surroundings at the rhabdom tip, as in ordinary apposition eyes, we called this system afocal apposition.[8] From an evolutionary point of view, this provided the link we were looking for with the superposition system in moths. By increasing the magnification of what are, in both moths and butterflies, double lens-cylinders, butterflies have reduced the superposition pupil from tens or hundreds of facets down to a single facet. It is not difficult to imagine how the butterfly system could have come about over time by a progressive increase in the optical power of the lower lens-cylinder.

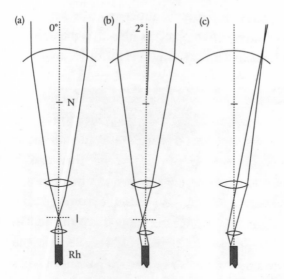

Fig. 30. The two-lens-cylinder optics of a butterfly ommatidium. The lens cylinders are drawn here as conventional lenses. a) A beam entering parallel to the axis is reduced in diameter so that it 'fits' the rhabdom. b) A beam 2° off-axis makes an angle of 12.8° when it enters the rhabdom, which just allows it to be trapped by total internal reflection. c) The rhabdom tip is imaged onto the cornea, making it possible to see the waveguide modes in the rhabdom.

Butterfly eyes have another feature that makes them wonderful study objects. In most butterflies there is a mirror at the base of each rhabdom, made from thin layers of chitin and air. These mirrors reflect characteristic colours—orange and green in *Heteronympha*. When the eye is lit from the same direction as the viewer, there is a bright-coloured spot in the centre of the pseudopupil (see Fig. 22) formed by light that has been to the bottom of the rhabdom, is reflected back, and emerges through the same facet that it entered. This bright spot fades within a few seconds—which is why there is no bright spot in the pseudopupil in Figure 22. This is a (true) pupil mechanism. Dark-absorbing pigment around the rhabdom is summoned by the receptors' response to light, and when it reaches the outer wall of the rhabdom, it 'bleeds' light out, so that very little light reaches the bottom of the rhabdom, and even less gets back out. This protects the sensitive photopigment from too much light—from the sun, for example.

With a suitable optical arrangement, it is possible to illuminate, and view, a much wider region of the eye, so that the reflected glow appears over a large number of facets. If an object is placed in the light path, it is imaged onto the rhabdoms in a mosaic way, and its form can be seen in the pattern

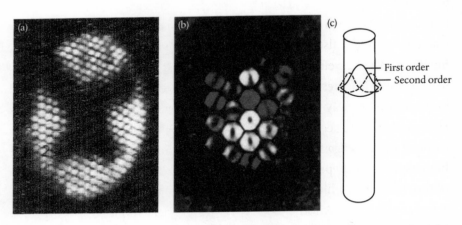

Fig. 31. a) Erect apposition image of a rather badly drawn butterfly, seen against the reflected glow from a patch of illuminated facets. b) Second-order waveguide modes in the rhabdoms of *Vanessa itea*, photographed at the corneal surface. c) Diagram of the distribution of light in a narrow waveguide, showing first- and second-order modes. Note that some of the light travels outside the rhabdom.

reflected back out of the ommatidia in the illuminated facet patch (Fig. 31). Butterflies are probably the only insects in which an overall erect apposition image, like that in Figure 15, can actually be seen.

These eyes have one final optical treat. Once again this is provided by the mirror at the base of each ommatidium, which makes it possible to see something of what happens inside the rhabdom itself. When light is confined in narrow cylinders, it behaves rather like sound in organ pipes: it sets up standing waves known as 'waveguide modes' (Fig. 31 right). Again, like resonant sound waves, the light in these stable modes forms distinct patterns, which depend on the thickness of the cylinder. The simplest pattern has a bell-shaped cross-section with a maximum in the centre, but as the cylinder thickness increases, modes with double peaks appear, and then modes with increasing numbers of peaks, until eventually the mode structure becomes unrecognizable. The peculiar optics of butterfly eyes makes it possible to study the mode in the rhabdoms simply by looking at the patterns in the corneal facets. The reason this works is that the reflected light emerging from the rhabdom tips is imaged, with its pattern much magnified, onto the corneal surface (Fig. 30c). Figure 31 shows mode patterns photographed in the cornea of a *Heteronympha* eye. They are mostly second-order (bilobed) modes with a central minimum. The first-order modes with a central maximum are also present, but they have been removed using polarizing filters. Besides being a beautiful phenomenon, the mode patterns can be used to study features of the rhabdoms themselves. For example, small blue butterflies show only the first-order mode; larger diurnal butterflies such as *Heteronympha* have second-order modes; and crepuscular butterflies such as tropical owls (*Caligo* spp.) and the Australian butterfly *Melanitis leda* have higher-order mode patterns. These differences are related to the diameters of the rhabdoms, which are wider and so trap more light, in animals that fly at lower light levels. Another intriguing feature of the modes in *Heteronympha* is that when the eye is illuminated, the pupil mechanism cuts out the second-order modes first, leaving the narrower first-order mode. This happens because the second-order modes, being wider, are more easily absorbed by the pupil—the screening pigment in contact with the rhabdom wall. This change also reduces the acceptance angle of the ommatidium, which means that resolution is improved in daylight, but more light is absorbed by the rhabdoms in dimmer conditions.

In spite of all the astonishing optical virtuosity that insect eyes display, there remains a mystery: why did insects not abandon them for something better? The diffraction limit imposed by the small lens systems keeps their resolution low, and single-chambered eyes of a similar size would give a much better performance. The eyes of some jumping spiders, for example, provide resolution at least six times better than the most acute dragonfly, and with much smaller eyes. There *are* single-chambered eyes in insects. Many insect larvae have them, and use them effectively in predation. Most flying insects have three small single-lens eyes on the top of the head (dorsal ocelli), and these act as horizon detectors during flight. One feels that at some stage in evolution, they could have taken over the functions of the compound eyes, but this did not happen. Perhaps the reorganization of the brain it would have required was insurmountable. Commenting on the persistence of compound eyes, Dan-Eric Nilsson remarked: 'It is only a small exaggeration to say that evolution seems to be fighting a desperate battle to improve a basically disastrous design.'

3

VISION IN THE OCEAN

The first time I looked into a bucket of animals just brought up by a mid-water trawl I was astonished. I had expected everything to be some sort of grey, but far from it. Some animals were black, some red, and some transparent, but many were beautifully silvered and quite jewel-like. Many of the invertebrates, especially the smaller crustaceans, were still alive and kicking, but the fish were usually dead because the gas bladders with which they adjust their buoyancy had been ruptured by the 50-atmospheres decrease in pressure.

I was introduced to this remarkable environment when I was invited to take part in a series of four cruises on the Royal Research Ship *Discovery* (Fig. 32). These cruises, between 1976 and 1990, were all to a region off North Africa where upwelling currents made the marine life plentiful and diverse. They were organized by Peter Herring, a chief scientist at the National Institute for Oceanography, whose main interest was in bioluminescence. He invited me along because he knew I would be intrigued by the ways animals managed to see their food, their enemies, and each other in the dim conditions 500 metres down.

The scientific laboratories on the ship (Fig. 33) could be described as primitive by the standards of labs on shore, but they were fine for what I wanted to do, which was mainly microscopy, dissection, and observation. These labs were sociable places, with plenty of banter of a more or less nautical kind, but more importantly, there was always someone who knew the name of any animal that caught your interest. Trawling involved paying out kilometres of cable until the net reached the right depth, when it was opened by electronic gadgetry. Recreation involved mild gambling and a lot of gin and tonic, the former becoming much cheaper than the latter once the ship had

Fig. 32. Royal Research Ship RRS *Discovery*. *Discovery III* was built in Aberdeen, in 1962, for deep-sea studies. She was the third generation of ships with that name that followed the illustrious *Discovery* that took Captain Scott to the Antarctic. She was quite small (4370-tonnes displacement) and slow, cruising at 11 knots, but her diesel-electric motors enabled her to sail at a controlled speed of 2–3 knots, which was ideal for deep-water trawling. She has now been replaced by a new ship, *Discovery IV*.

reached international waters, the magic line being three degrees West of Ushant. There was an omnipresent aroma of diesel oil and chips.

The extraordinary appearance of mid-water creatures, which had so impressed me, only starts to make sense when you try to piece together their lifestyles. What really matters in these deep waters is the nature and distribution of the remaining light from the surface. For a start the light is blue; long-wavelength light from green to red, and short wavelength violet and ultraviolet light are all absorbed strongly by water, leaving only blue. One consequence of this is that red and black are interchangeable if an animal does not want to be seen. Even for blue light, absorption reduces the intensity by a factor of ten for every 70 metres, so by 700-metres depth the light levels are 10^{10} times lower than at the surface. The absolute threshold at which a human observer could just detect daylight would be at about 800 metres, when looking upwards. Useful vision, as we all know from

Fig. 33. The main scientific laboratory on the *Discovery* in 1976.

dark nights, requires much more light than this. Even for fish, with larger lenses than ours, this means that there is no usable daylight below 1000-metres depth. Even at mid-day the mid-waters around 500 metres are really a twilight zone, and vision has to be thought of in those terms. The key to understanding the appearance of animals at these depths is that they need to see others while remaining invisible themselves.

Another important aspect of the residual daylight is the way it varies with direction. Wave refraction at the sea surface, together with scattering, diffuse the light so that it becomes symmetrically distributed around the vertical (Fig. 34a). Downwards light is strongest, and light scattered back upwards is less than one hundredth as bright, with light from the side somewhere in between. The important point is that light making an angle of, say, 30° from the vertical has the same intensity in all directions. This provides an explanation for the silvering of the sides of many of the fishes. A vertical mirror placed in a light-field like this becomes invisible, because the light reflected from the mirror has the same intensity as the light *that would have come through it* (Fig. 34b). This means that if a fish can contrive to make its sides into flat mirrors it will become invisible from almost all directions—except from directly below: a fish still has a silhouette that blocks light. The trick for a fish is to make a flat mirror from its far-from-flat flanks. Some fish, such as hatchet fish, are remarkably flat-sided anyway, but this is unusual; most fish have convex sides, which, if silvered, will not provide camouflage: only a flat mirror will do this. Yet fish manage to achieve this seemingly impossible transformation. In the 1960s Eric Denton and Colin Nicol tackled the question of how this works.[1] They worked on silvery fish such as sprat and herring at the Marine Biological Association Lab in Plymouth, studying how the tiny reflecting plates within the scales of the fish are organized. They measured the angles at which light was reflected from different regions of the side of the fish and found that the reflecting plates are not in general in the same plane as the fish's surface. Instead they are tilted at angles to the body surface which get larger towards the upper and lower parts of the flank. The effect of this is to produce a *reflecting* surface which is more or less flat, and, provided the fish is swimming upright in the water, this surface becomes a good approximation to a vertical plane mirror (Fig. 34c). When swimming normally, in a shoal or alone, fish have a 'dorsal light reflex' which keeps the body upright, particularly at moderate depths in the sea,

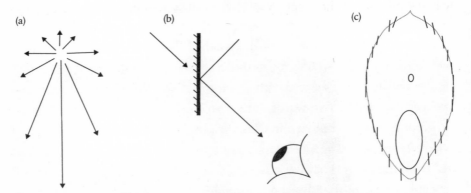

(a)   (b)   (c)

O

Fig. 34. Reflecting camouflage in the sea. a) Distribution of light around the vertical in the deep ocean. b) Light reflected from a mirror has the same intensity as light that would have passed through it. c) The vertical alignment of reflecting plates around a transverse section of a silvery fish such as a herring.

where the brightest light always comes from vertically above. This technique for turning a round surface into a functionally flat one really works. Divers sometimes report seeing shoals of small black dots, which turn out to be the pupils of the eyes of silvery fish whose bodies are invisible to them. Fish sometimes break this camouflage to signal to each other. It only requires a brief flick away from the vertical to produce a flash, which can be used to coordinate movements within a shoal (Fig. 35).

Reflective camouflage is very effective at disguising a fish from the side over a wide variety of angles, while disguise from above is simply a matter of making the dorsal surface black. But from below, these forms of camouflage will not work because the fish's silhouette means that it always blocks out some of the light from above and so becomes visible to an upwards-looking predator—of which there are many.

Many fish, crustaceans, and squid have a particularly ingenious method of overcoming this: they replace the light they block by creating their own substitute light using bioluminescence. Bioluminescence is extremely common in the deep sea. At depths between 200 and 1000 metres, its importance increases, and below 1000 metres, it is the only source of light. Bioluminescence results from the activation of a luciferin molecule—of which there are many varieties—by an enzyme, luciferase. A rather complicated swapping of chemical bonds in the luciferin molecule releases energy

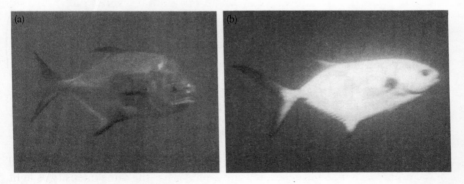

Fig. 35. A silvery game fish, *Trachynotus falcatus*. a) Upright in the underwater light field. b) Tilted, so that the reflecting camouflage is temporarily compromised.

in the form of light. The majority of luminescent animals have their own light-producing systems, but some others employ light-producing bacteria, which they maintain in special pouches.[2]

The uses of bioluminescence are extraordinary varied. A far from exhaustive list would include making flashes that deter predators, and a variant of this in which the organism—usually a crustacean or squid—emits a cloud of luminous material and then escapes into the dark. Some fish have a pair of large emitters on their heads facing forwards, and presumably, they use these as searchlights to find prey in the water ahead of them (Fig. 36). The light produced by photophores (light-producing organs) of most marine animals is blue: the colour of residual daylight, to which the eyes are most sensitive. However, one group, the dragon fish, or stomatioids, have photophores that produce red light and visual pigments in their eyes that can respond to it. At short range, this red light gives them a private channel of communication because nothing else can see it, and it has been suggested that this can also be used to break the camouflage of red or orange crustacean prey, which would appear black to normal blue-sensitive predator species.[3] Many angler fish use luminous lures on projections above the mouth, and they devour animals that come to investigate the light. In other animals, bioluminescence is used in sexual display, with the two sexes having different patterns of photophores. Of all the uses of bioluminescence in the sea probably the most common is for camouflage, in which downwards-directed photophores are used to make an animal's silhouette invisible against the dim light from above.

This form of disguise—counter-illumination—is a tricky business. It is obviously important that the intensity of the luminescence matches that of the residual skylight. This is far from easy to achieve, particularly as the light is emitted in a direction that the eyes cannot see. In hatchet fish this problem is partly overcome by having a very small photophore pointing *into* the eye (Fig. 37); the animal sets the intensity of the ventral photophores by matching the output of this photophore to the light from above. Even then there is no way of checking that the signal the eye provides to the ventral photophores is accurately calibrated. The hatchet fish arrangement does solve one other problem: this is the change in light intensity over the dawn and dusk periods, when the downwards light can change by a factor of a million. Fish, squid, and shrimp have all been shown to be capable of changing the intensity of their bioluminescence over a range of at least 10^4, in line with the light to which they are exposed. This implies that the supply of energy to the photophores is under subtle control.

According to Peter Herring,[2] the counter-illumination principle was used with partial success in World War II to camouflage torpedo bombers. Lights were mounted under the wings and fuselage and a photocell-controlled feedback system matched their output to the light of the sky above. This did

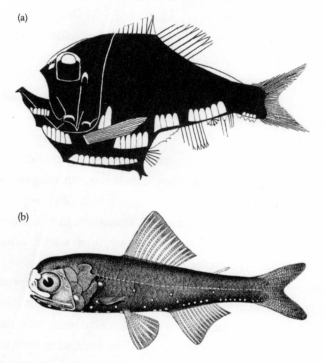

Fig. 36. Patterns of photphores on a) a hatchet fish (*Argyropelecus aculeatus*) and b) a headlamp fish (*Diaphus effulgens*). The hatchet fish has three rows of downwards-directed photophores on its lower flanks, and an upwards-pointing eye. The headlamp fish has a pair of large forwards-pointing photophores between its eyes, to light up the water in front. It also has 'camouflage' photophores—the small white spots on its ventral flanks—and a silvery band along the body.

(a)

(b)

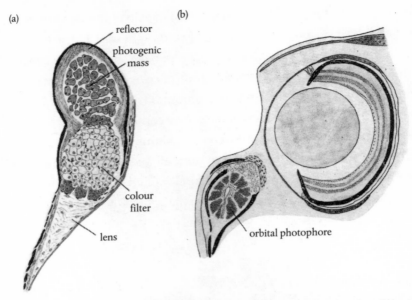

reflector

photogenic
mass

colour
filter

lens

orbital photophore

Fig. 37. a) Section through a 'camouflage' photophore of a hatchet-fish. The 'photogenic mass' is the location of the luminous pigment. b) Photophore directed into the eye in *Cyclothone braueri*.

greatly decrease the range at which the plane was visible from submarines. The project was prematurely shelved because the development of radar reduced its usefulness.

Both the fish and the plane examples demonstrate that the way counter-illumination is achieved depends on the range at which camouflage is intended to be effective. In fish the photophores produce what are almost point sources of light, which from a distance blend to give an average that is a match for the downwards light (Fig. 38). Close up, however, they will look like small, bright lights. Much will depend on how well the predators can resolve detail, which will not be great at mid-water light levels. But it is clear that the efficacy of counter-illumination will work best at longer distances, and at closer distances, many smaller photophores will be more effective than a few large ones.

The intensity of light emission is not the only variable that needs to be considered. The colour of the light must match that of residual daylight,

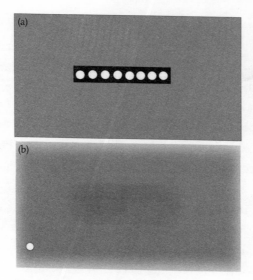

Fig. 38. A demonstration of the effectiveness of counter-illumination. a) shows a dark silhouette against a dim background, with bright 'photophores' added. Close up this pattern is highly visible. b) is the same, but it has been blurred to imitate the effect of viewing the pattern from a distance. The blurring is done by moving a circle (shown in white) across a), averaging the light within the circle, and reconstructing the scene from these average values.[4] The silhouette-photophore pattern almost disappears, and if the circle is made 10% wider it disappears completely. The distance at which this will happen depends on the resolution of predator's eye, but if it can only resolve 1°, which is a reasonable guess, this disappearance will occur at a distance of about 60 cm. Getting the ratio of dark silhouette to photophore output is crucial—and hard to achieve. If the ratio is wrong, the silhouette will continue to appear darker or lighter than the background.

and many photophores have filters built into them to ensure an appropriate match. The angular distribution is important too; the better the match between the spread of light from the photophores and that in the oceanic light, the more effective the camouflage will be. Some photophores have an elaborate arrangement of reflectors and baffles to achieve an appropriate light distribution (Fig. 37).

Many mid-water animals, particularly shrimp and squid, are more or less transparent, which of course is the perfect disguise. However, there are certain organs—particularly the eye and the contents of the digestive system—

that cannot be made transparent, and usually these parts have photophores on their ventral surfaces, and often silvering on their sides as well. Orientation is important, because counter-illumination only works if the photophores point downwards, and many animals change their attitude in the water as they swim. Squid tend to stabilize their eye orientation in relation to gravity, so their eyes' photophores remain downwards directed. Shrimp and krill (euphausiids) have the ability to rotate their photophores in the pitch plane, so that their direction is unaffected by the animals' undulating swimming movements.

The need for such elaborate camouflage strategies becomes clear when one remembers that the production of new material from plant life is confined to the first few tens of metres at the ocean surface, and at mid-water depths there is essentially nothing to eat but other animals and occasional droppings from animals in the upper waters. Unlike life on land, mid-water animals have nowhere to hide, and the only way to avoid being seen is to look as much as possible like the light distribution of the background. None of these animals can actually see themselves as others would see them, and so have no opportunity of tweaking the details of their camouflage to minimize their visibility to others. This means that all these beautiful mechanisms have come about, and been continually refined, by natural selection—the survival of the least visible.

Eyes

Most mid-water animals are carnivores, looking for other animals to eat, and it is not surprising that most of them have large eyes. In many cases these are directed upwards to detect just the kinds of silhouette that both they and their prey are anxious to avoid. This trend towards upwards-pointing eyes is not confined to fish; some squid, and crustaceans with compound eyes—euphausiids (krill) and hyperiid amphipods (crustaceans distantly related to sandhoppers)—show the same adaptations. The eyes are often double, with a large eye pointing upwards and a smaller eye directed obliquely

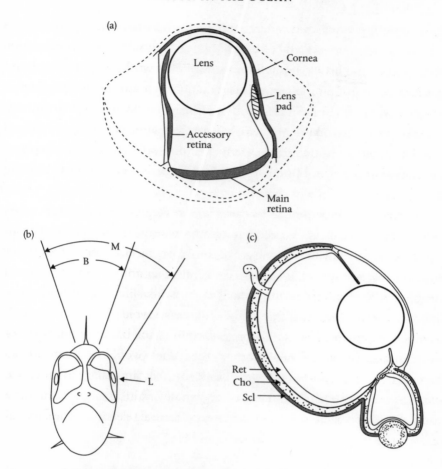

Fig. 39. Tubular and double eyes of mid-water fish. a) *Scopelarchus* section showing how each 'tubular' eyes is a cut-down hemisphere, where the main retina has an upwards field of about 60°. The upper part of the accessory retina views light projected through the lens from the lens pad, which acts as a light-guide. b) Monocular (M) and binocular (B) fields of *Scopelarchus*. c) Double eye of *Bathylychnops exilis*. L, lens pad; Ret, retina; Cho, choroid; Scl, sclera.[5]

downwards. The function of these secondary eyes, which look into the darkness of the abyss, is more mysterious. Their most likely function is to detect creatures that emit light for purposes other than camouflage.

Figure 39a shows the eyes of a mid-water fish, *Scopelarchus*. Most fish eyes are more or less hemi-spherical, but in this genus, and in many other predators from deeper regions, the eyes are tubular. The reason for this is that they do not need the full 180° field of view of the eye of a surface fish, because surface light that is bright enough for useful vision is confined to a relatively narrow cone around the vertical (Fig. 34a). These eyes are really cut-down spheres, rather than cylinders, with a retina (the 'main retina' in Fig. 39a) viewing an angle of about 60°, which covers the region above the fish where silhouettes might be detectable. Both eyes of Scopelarchus point in the same upwards direction, and have a binocular field of view of about 30°, where their ability to detect weakly contrasting objects will be greatest (Fig. 39b). The hatchet fish in Figure 36 has eyes with a similar tubular shape, and a similar binocular field. Scopelarchus eyes have another strange feature, an opalescent 'lens pad', which consists of multiple narrow plates of refracting material that behave as light-guides, and collect light from obliquely downwards. This is projected through the lens onto the top of the 'accessory retina' (Fig. 39a). It seems that this is one of a number of ways that fish manage to image bioluminescent light from below. In Scopelarchus this light will be poorly focussed at best, but in other cases, the optical arrangements are much easier to understand. Bathylychnops exilis has a second, downwards-pointing eye with a lens and retina of its own (Fig. 39c). This eye is not just a small duplicate of the main eye: its lens is formed from a swelling of the sclera—the wall of the eye—rather than being an independent structure like the lens of the main eye.

Perhaps the strangest example of a secondary eye is found in the spook-fish, Dolichopteryx longipes (Fig. 40). According to Ron Douglas, this fish was left in the bucket from the trawl, after the other scientists had had their feeding frenzy, because nobody knew what it was. Seen from above the eyes are clearly double structures, but only the main eyes have lenses. Instead the secondary eyes have a particularly clever mirror system that allows them to focus light from below over an angle of close to 50°. In the downwards-pointing region of the secondary eye, there is a window which admits light onto a mirror on the inner wall of the eye, and this mirror forms an image on a laterally placed retina. This is no ordinary mirror. No single reflecting surface could produce a flat image over such a wide angle. Instead, as its

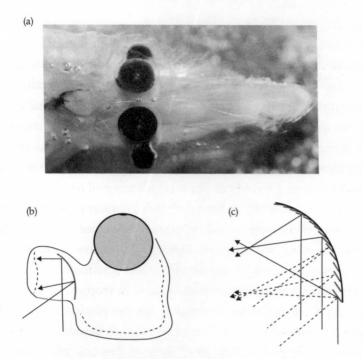

Fig. 40. Reflecting eyes of the spookfish, *Dolichopteryx longipes*. a) View of the head from above, showing the double eyes on each side. b) Diagrammatic section, showing the main eye with a lens, and the secondary eye, with a mirror (grey) and retina (dots). c) The stepped mirror, showing how it collects light from both directly below (solid lines) and from 45° to the vertical (dashed) to form a well-resolved image in the plane of the retina.

discoverers found in 2009, it is made of a series of stepped reflecting plates that make increasing angles with the surface behind them, a little like the plates on the sides of a silvery fish (Fig. 34c). This angling of the plates makes it possible to image all the light reaching the mirror onto the more or less flat plane of the retina. This way of forming an image is unique to this group of fishes, and it is the only example of an image-forming mirror in the vertebrates.[6]

The secondary eyes of these double-eyed fish are all smaller than the main eye, and certainly in the case of the *Scopelarchus* lens pad, they produce rather indifferent images. This seems paradoxical because they are looking into an even dimmer direction than the main eyes, and one might expect them to be large and sensitive. It turns out, however, that detecting a luminescent flash against a dark background is a very much simpler task than detecting a low-contrast silhouette against a dim background. A flash is unambiguous, and it doesn't matter much whether the photons from it are well-focussed, so long as they are *caught*. A detector is what is needed, rather than a well-resolved image.

I spent much of my first two *Discovery* cruises studying the compound eyes of carnivorous crustaceans. The relationships of these groups are shown in Figure 45. Some of the eyes are extraordinary, and quite unlike anything seen on land or even in shallow water. In particular, the enormous transparent eyes of hyperiid amphipods took me by surprise, both for their beauty and the complexity of their design. When preserved they look quite drab, and few people have seen them in their living glory: I feel privileged to have had the rare opportunity. Optically they function as ordinary apposition eyes, but on first acquaintance, you would hardly have guessed this. These animals are quite common in mid-water trawls, from depths of 100–800 metres, and they survive well at the surface. Like many fish from these depths, they often have double eyes, and no doubt for the same reasons. My favourite is the 'Diogenes shrimp' *Phronima sedentaria* (Fig. 41a). It is quite a large animal, 2–3 cm long, and lives in a transparent barrel that it makes by hollowing out the inside of a salp, a kind of oceanic sea-squirt. (The shrimp's common name refers to the philosopher Diogenes, who famously spent some time living in a barrel.) The barrel is protective, and the females keep their brood inside it; the adults sometimes get out and push the barrel from behind, like a strange baby carriage. The whole of the top of the shrimp's head is covered by the surface of a pair of huge eyes, which look like a delicate parasol in the picture. Each has about 400 facetlenses. These can be seen in Figure 41b, together with a very large pair of black pseudopupils—the regions of the eyes looking into one direction in space. Like other pseudopupils (see Chapter 2), these move across the eye surface as the viewing direction changes. These eyes have other extraordinary features: the facets are huge. At nearly 0.2 mm in diameter, they are ten times wider than those of a typical insect, and so have one hundred times the light-collecting power. The field of view of the two eyes is very narrow, barely 10° across.[7] The retinas the eyes supply are the inner of the two pairs

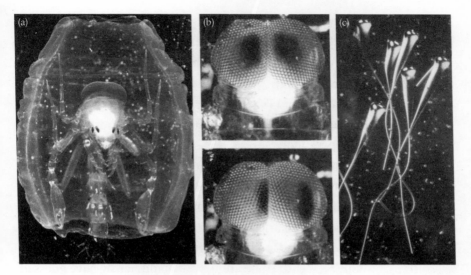

Fig. 41. *Phronima sedentaria.* a) Living *Phronima* in its barrel. The four black dots are the retinas of the two eyes. b) Views of the upper-eye surface, showing the lenses and the pseudopupils from two angles. c) Lenses and light-guides from the upper eyes.

of dark dots at the bottom of the head in Figure 41a, and the lenses are connected to these by transparent light-guides 5 mm long, probably the longest light-guides in any animal (Fig. 41c). The reason for having such a condensed retina seems to be simply to keep it small, so that it barely disturbs the transparency, and hence invisibility, of the animal as a whole.

The outer pair of dots is the retinas of the lateral eyes, which can just be seen as teardrop-like structures surrounding each retina. The much smaller size of these eyes is reflected in their low resolution: the inter-ommatidial angle is about 10°, compared with less than 0.5° in the upper eyes. As in double-eyed fish, the assumption is that the upper eyes search the residual light from the surface for silhouettes of prey animals, which can be either other crustaceans, or more or less transparent creatures—jellyfish medusae or the salps from which *Phronima* make their barrels. The lateral eyes may be concerned with low-resolution all-round vision or for detecting luminescent prey, much as in fish.

Other mid-water hyperiids have double eyes not so different from those of *Phronima.* In *Phrosina, Brachyscelus,* and *Parapronoe* the eyes are divided into upper and lower halves, but are not separate structures as in *Phronima*. They are transparent hemispheres that look rather like the front end of an observation helicopter. In all of them, the upper eye has higher resolution and a narrower field of view than the lower has. In hyperiids that live nearer the surface, such as *Lestrigonus* and *Thamneus*, the distinction between upper and lower eyes disappears, presumably because there is adequate light in all directions. The deepest living of all the hyperiids, *Cystisoma*, is a large (20-cm) totally transparent and rather floppy animal. It only has upper eyes, whose two retinas stretch like a slightly orange curtain across the whole of the top of the head. It is so dark at 800 metres that the eye has no need for screening pigment to cut out stray light.

An eye that takes the prize for sheer strangeness is that of *Streetsia challengeri* (Fig. 42). The back end of the animal is fairly ordinary and shrimp-like, but the head end is unlike anything else. The eye is sausage-shaped, about 7 mm long, and in front of that is a spike of similar length. The retina is the dark body enclosed in the lower part of the eye, and like other midwater amphipods, the eye is asymmetric, with most of the surface devoted to the upper field. The strange shape of the eye means that the resolution in the fore-and-aft direction is much better than in the transverse.

Like the eye itself, the field of view is cylindrical, but it does not extend to the forwards direction, where the spike is. It is not easy to see how all this might work. The animals are quite speedy, and one can imagine them detecting a

Fig. 42. The head end of the hyperiid amphipod *Streetsia challengeri*. The eye is the sausage-shaped structure on the right, and the retina is the dark structure in the lower part of the eye. To the left in front of the eye is a 7-mm spike. The eye has no forwards field of view, so the animal cannot see what the spike will run into.

soft-bodied creature overhead before dashing up and spearing it. But would a forwards view not help here? A deeply enigmatic animal.

Another crustacean group in which the deep-water species have double eyes are the euphausiids or krill. They are a sister group to the better known, and tastier, decapod shrimps (see Fig. 45). The big Antarctic species, *Euphausia superba*, is up to 6 cm long. It lives near the surface and has spherical eyes. These shrimp-like animals often live in huge schools, feeding on other small crustaceans, and are themselves preyed on by baleen whales, which simply filter them out of the water. The compound eyes in all euphausiids are of the superposition type, with the same basic clear-zone design and refracting optical system as those of moths and fireflies (Chapter 2). One has to assume that this is a case of independent convergent evolution, like the eyes of fish and squid. All species have downwards-pointing photophores to disguise

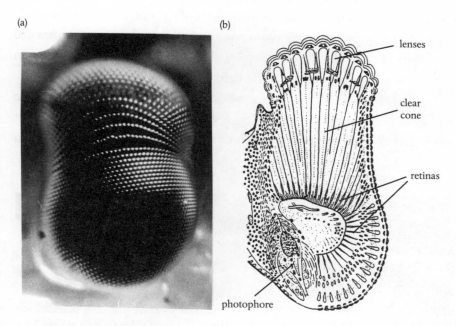

Fig. 43. Double superposition eyes of mid-water euphausiids. a) *Nematoscelis megalops*. The eye is 2.5 mm high. b) Section through the eye of *Stylocheiron suhmii*. The upper eye has a narrower field and higher resolution than the lower eye. The eye has a downwards-pointing photophore to disguise its dark silhouette.

the eyes and other non-transparent parts of the body (*Phausis* is a Greek word meaning *lighting* or *illumination*; there is also a firefly genus *Phausis*).

Descending from the surface through the mid-water layers one finds the same trends in eye shape in different species of euphausiid as in the hyperiid amphipods. At the surface the eyes are spherical; then a small higher-resolution upper part develops (Fig. 43). This gets larger and the lower part gets smaller until, in *Nematobrachion boopis*, the eye is almost entirely taken over by the upwards-pointing region. These trends, and the counter-illumination disguise of opaque features, imply that all the mid-water crustaceans share with fish the same basic strategy of looking upwards for food, while making sure they are not seen from below.

Before leaving the mid-waters of the ocean, I should mention one unusual squid genus, *Histioteuthis*, also known as the cock-eyed squid. Most animals mentioned so far, fish or crustacean, have had two eyes each with two parts, one directed upwards and a smaller one pointing downwards. *Histioteuthis* is

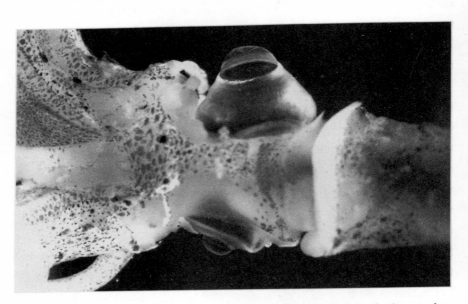

Fig. 44. The 'cock-eyed squid', *Histioteuthis* (probably *bonellii*), partly dissected to show the shapes of the eyes. One eye is elongated with a reduced field as compared with the other smaller eye. In life the larger eye is directed towards the sea surface, and the smaller eye downwards.

similar, except that its two eyes each seem to be specialized to do only one of the two tasks: scrutiny of the light from above and lower resolution surveillance of the dark below. Figure 44 shows a specimen, partly dissected to show the profiles of the two eyes. The larger eye is conical in shape, with a field of view of about 80°, while the smaller eye is a hemisphere with a much wider field of view. The lens of the larger eye has twice the diameter of the smaller, and twice the focal length. Little is known about the behaviour of these animals in the sea, but it is hard to resist the idea that the division of labour between the two eyes in *Histioteuthis* is the same as that between the principal and secondary eyes of fish (Fig. 39c).

Shrimp and Mirrors

Most of the crustaceans worth eating are decapods, the class that includes shrimps, prawns, crayfish, lobsters, and crabs. Typically, they have an eight-segment thorax and a six-segment abdomen, and, unsurprisingly, ten legs. They make up about three quarters of all known crustacean species. Their relations to other crustaceans are shown in Figure 45.

Until 1975 it was not at all clear how the eyes of decapod shrimps and lobsters worked. They are certainly superposition eyes, because they have a clear zone between the optics and the receptors, but the optical elements that are supposed to direct light to a focus did not seem capable of doing this. The lenses in the eyes of krill are hard, bullet-shaped lens cylinders with an internal gradient in refractive index, just as in moths or glow-worms. Shrimp 'lenses', however, are four-sided truncated pyramids made of more or less homogeneous jelly (Fig. 46c), and this is hardly a good basis for any kind of optical system. The square geometry of the facets also set these eyes apart from almost all other compound eyes (Fig. 46a).

On the 1976 cruise of the *Discovery*, I thought I had found the answer.[8] Some of the mid-water shrimp have eyes with less dark screening pigment than in surface-living decapods, and so one can get a better view of what these enigmatic optical elements really look like. When dissecting the eye of

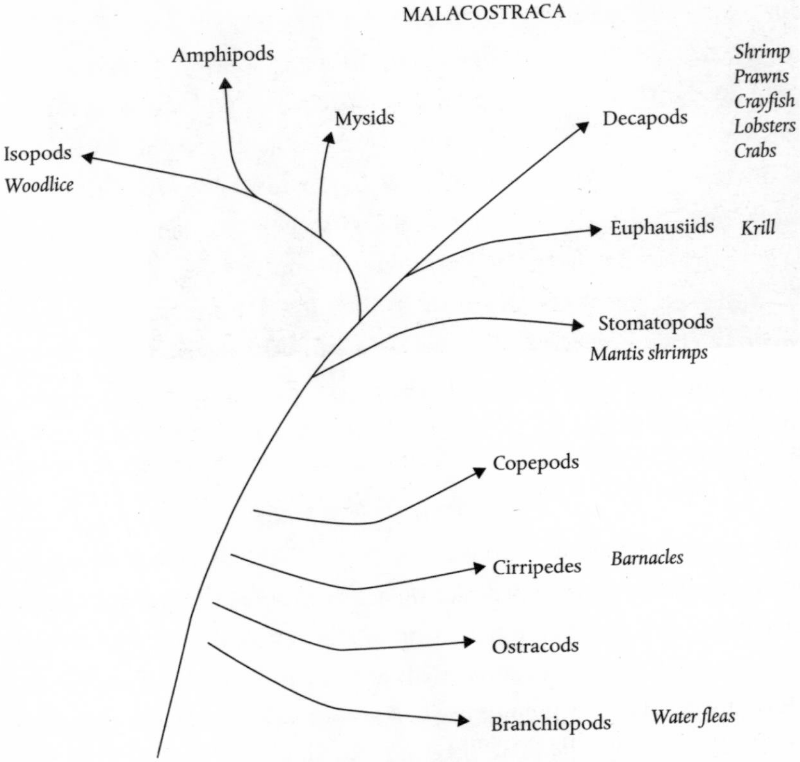

MALACOSTRACA

Fig. 45. Simplified family tree of the Crustacea, characterized as arthropods with two pairs of antennae. The upper-six orders make up the class Malacostraca, which includes most of the larger crustaceans. The remaining four classes used to be grouped as the Entomostraca, but their relationships are too obscure to justify this.

one of these shrimps, *Oplophorus spinosus*, I could see that the jelly pyramids were lined with a greenish-white reflecting material. They were not just jelly, they were *silvered* jelly. The fact that the sides of the elements were not round, but flat, meant that they were plane mirrors. That was it. A few minutes with a ruler and protractor and the job was done. It was clear that a series of mirrors, at right angles to the eye surface, could form a superposition image (Fig. 47). This was a new kind of eye.

This was very exciting. I wrote it up and sent the paper to *Nature*, who published it. Shortly afterwards I had a letter from my friend Kuno Kirschfeld in Tübingen, pointing out that someone else had reached similar conclusions,

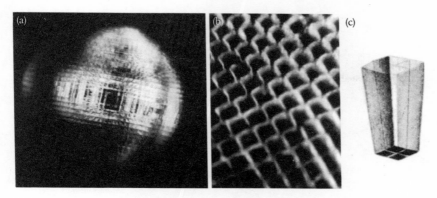

Fig. 46. Reflecting superposition. a) The eye of a shrimp, *Palaemonetes varians*, showing the silvery appearance and the square facet geometry. b) Surface of the eye of a living crayfish, showing the tips of the square mirror boxes. c) One of the truncated jelly pyramids in a shrimp eye drawn by Hermann Grenacher in 1879.

working on eyes of crayfish (Fig. 46b). Klaus Vogt, also from Tübingen, had published a short note in *Naturforschung* the year before (1975). Here is what he wrote:

> Rays from an object point entering through different facets are superimposed not by refracting systems as in other superposition eyes but by a radial arrangement of orthogonal reflecting planes which are formed by the sides of the crystalline cones and the purine layers surrounding them.

Touché. I wrote to Kirschfeld and Vogt and they were quite happy with the situation; it does no harm for the same discovery to be made twice independently; in fact, it reinforces the probability of it being right! It was clear that Vogt was going to continue to work on the crayfish eye, and I was happy not to pursue the shrimp work further. I did subsequently find out that the optical elements in a shrimp eye (the mirror boxes) had been drawn with exemplary accuracy in 1879 by the great anatomist Hermann Grenacher (Fig. 46c). He didn't comment on their shape, nor notice the silvering, presumably because this had long since vanished in the preserving process. Sigmund Exner, who got so many other things right in his 1891 book, had little to say either. He noticed that the facets of the prawn *Palaemon* were square, but he believed the structures beneath the facets were lens cylinders, much as in glow-worm eyes.

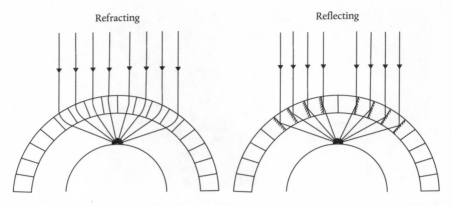

Fig. 47. Superposition images can be formed by refracting lens cylinders (glow-worms, moths, and krill), and by radially oriented mirrors (decapod crustaceans).

Vogt went on to explain the significance of the square shape of the mirror boxes.[10] It needs an explanation because two-dimensional arrays of most structures form hexagons, because this geometry gives the greatest packing density. Honeycombs and insect eyes are examples. Squares look somehow unnatural. The answer for their existence in decapod crustaceans is that two sides of each mirror box form a *corner reflector*, and such reflectors have the property that they always return light in the direction it came from (Fig. 48a). You meet them sometimes in clothes shops, hairdressers, or toilets in upmarket hotels. When you look at them from different directions, they always reflect your image back. The reason for this is simple: from whatever direction you look, the returning rays have been reflected by the two surfaces through an angle that adds up to 180°. A version of the corner reflector, but with three reflecting surfaces at right angles, is commonly used as a radar beacon on ships and buoys, because it too reflects radar waves back to their source.

The role of corner reflectors in shrimps' eyes is subtle and best explained by imagining an alternative. The cross-section shown in Figure 47 could be of a number of structures, but suppose it were a section through a series of continuous circular reflecting strips, all angled towards a common centre at C. One such strip is illustrated on the left of Figure 48c. Such a structure has a single axis, and while it would form a focus at F for rays parallel to that axis, rays from other directions would go all over the place, and there would

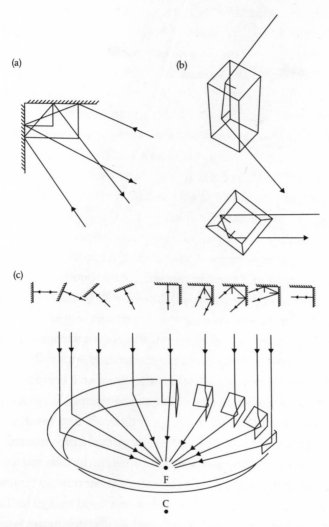

Fig. 48. The mirror boxes in shrimp eyes behave as corner reflectors. a) Simple corner reflector showing that all incoming rays are reflected back through 180°. b) Three-dimensional path of a ray through a mirror box in a shrimp eye. Seen from above the box, the ray emerges parallel to the path it entered, as in a). c) Showing how a continuous reflecting strip can be replaced by an array of corner reflectors, forming an image at F. Insets show ray paths seen from above. C is the eye's centre of curvature.

be no focus. Replacing the single strips with an array of corner reflectors solves this problem (Fig. 48c right). Because corner reflectors always reflect light *as though* they were mirrors at right angles to the incoming rays, such an array no longer has an axis, and this allows rays to be properly focussed over a wide angle—potentially 360°. This will only work if most rays entering the eye are reflected from two surfaces in each mirror box (Fig. 48b). Three is too many, so the depth of each box is important: it needs to be two to three times the facet width, as in Grenacher's drawing (Fig. 46c).

There is an interesting sequel to this story. In 1978 I wrote an article for *Scientific American* about animal eyes with mirror optics. This was read by Roger Angel, an astronomer in the Steward Observatory in Tucson, who saw my diagram of the shrimp/crayfish/lobster eye and concluded that this optical design would work for X-rays. X-rays are problematic because they cannot be refracted, and they do not reflect at normal incidence—at right angles to a mirror surface. They will, however, reflect at very low angles— 'grazing' incidence—which is what the shrimp eye provides. Angel went on to design a wide-angle X-ray telescope, and although that design was not funded at the time, subsequent versions have been produced. Another application has been as a way of producing a parallel X-ray beam, effectively by putting an X-ray source at F in Figure 48c and using the optics in reverse. The value of this is in producing microcircuits, which are etched by X-rays rather than light, as the shorter wavelengths provide higher resolution.

Another version of the superposition principle was discovered by Dan-Eric Nilsson in 1988.[9] I had wondered for some time whether it was possible to combine lenses and mirrors in a compound eye optical system, and concluded that it wasn't. This turned out to be just a failure of my imagination. Nilsson had been studying crabs, which, like the shrimps and crayfish, are decapod crustaceans but have in general retained into adulthood the apposition eyes that all decapods have as larvae. However, some, such as the swimming crab *Macropipus*, have an arrangement that allows them to be both apposition and superposition eyes at the same time. The facets contain lenses which produce images near the bottom of each crystalline cone, as in an apposition eye, but the central part of the image does not

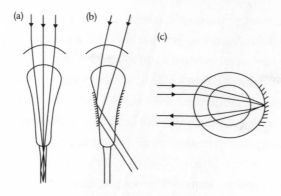

Fig. 49. Parabolic superposition. a) Optical structure as an apposition element. b) Optical structure as a superposition element in which the parabolic reflecting sides of the crystalline cone form a parallel beam from the partially focussed incoming beam. c) The same from above.

enter a rhabdom directly, as it would in a bee's eye, but passes into a light-guide which crosses a wide, clear zone before finally reaching the receptors. Rays that are more oblique do not enter the light-guide, but are reflected from the cylindrical walls of the crystalline cone, and emerge near the base of the cone on the same side as the incoming rays, as in other kinds of super-position eye (Fig. 47). The trick here is in the shape of the crystalline cones; these have a parabolic profile that counteracts the initial focussing of the rays by the facet-lenses (Fig. 49). In this way the converging rays emerge from the cones as a parallel beam, which crosses the clear zone to contrib-ute to a superposition image. This arrangement, which Nilsson has called 'parabolic superposition' is, as far as I know, the last form of image forma-tion to be described in a compound eye.

Biological Mirrors

Mirrors have been mentioned many times: in the eyes of scallops (Chapter 1), dragonfish, and crustaceans, and as camouflage in the reflecting scales of fish. They also form tapeta (*tapetum* means carpet or tapestry in Latin); these are the mirrors at the back of the eyes of many animals whose function is not to form an image but to return light back through the retina to double the chances of a rhodopsin molecule absorbing a photon. Cats' eyes are the obvious example, but tapeta are found in many other nocturnal vertebrates, as well as in moths, butterflies, spiders, and decapod crustaceans. Biological mirrors are often brightly coloured, and this makes them useful in advertisement. The wings of *Morpho* butterflies and the tail feathers of peacocks are famous examples where both the shininess and colour of such reflectors are exploited.

None of the mirrors just mentioned are metallic. There are several ways that biological materials can be used to make reflectors,[10] but the most common is the use of layers of film-like material. These films are each so thin that light waves reflected from the top and bottom surface of each film interfere with each other. This is the same phenomenon as seen in soap bubbles and oil films. Light can behave as a particle (photon), a ray, or a wave, but for our present purposes, it is the wave description that is relevant. If light waves from two sources are in step, so that the peaks and troughs of each wave coincide, they will add together, and the energy of the resulting wave will be enhanced. This is *constructive interference*. If, on the other hand, the waves are out of step, so that peaks coincide with troughs, the reflected waves destroy each other—*destructive interference*. A thin transparent film, such as a soap bubble, will appear bright if the light reflected from the lower surface has been through a whole wavelength in its double passage through the film, before meeting the wave reflected from the upper surface. The two waves will then be 'in phase'. On the other hand, if there is a half-wavelength difference the waves will be out of phase, and the film will appear dark. As white light contains all wavelengths, some will be reflected in phase and some out of phase, and so the film will appear coloured.

One might think from this that a film would need to be a half wavelength thick for the double passage through it to produce constructive interference for light reflected from both surfaces. There is, sadly, a catch. Owing to a piece of physics that is particularly hard to understand, light reflected from a surface separating a low-refractive-index material, such as air, from a higher index material, such as a soap film, automatically has a half-wavelength phase shift.

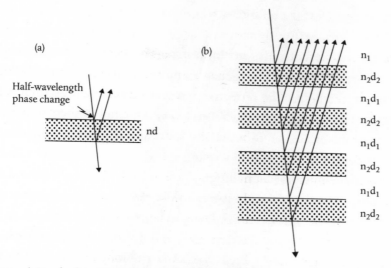

Fig. 50. a) Single thin film, such as a soap bubble. Because of the half-wavelength change at the low-to-high but not the high-to-low boundary, the film needs to have an optical thickness (refractive index **n** x actual thickness **d**) of a quarter-wavelength for the two reflected waves to interfere constructively. b) A multiple film, where all the high-index layers ($\mathbf{n_2 d_2}$) and the spaces between them ($\mathbf{n_1 d_1}$) have an optical thickness of a quarter-wavelength.

This does not happen at the other surface (film to air). The consequence is that the double passage through the film needs to be a half wavelength for constructive interference to occur, and so the film itself only needs to be a quarter-wavelength thick (Fig. 50a). If you look carefully at a soap bubble, you can see that the thinnest part is actually non-reflecting—black—because the two surfaces are almost touching: there is only the obligatory half-wavelength difference between the reflected waves, and this causes the two reflections to cancel. As the film thickens to a quarter wavelength, there is a broad bright area that is nearly white, followed by a succession of bands of strong colours, until these merge to give white again as the film gets thicker still. This sequence of colour changes was discovered by Isaac Newton, and it still bears his name—Newton's Series. In spite of this, Newton did not like the wave theory of light proposed by Christiaan Huygens; he preferred 'corpuscles'.[11]

A single quarter-wavelength-thick film only reflects a few percent of the incident light. In real biological reflectors this reflectance is increased by adding more films, all a quarter-wavelength thick and separated by quarter-wavelength spaces (Fig. 50b). In such a stack, the reflectance increases dramatically, so that with only ten film and space pairs, the reflectance reaches 99%; these mirrors are as effective as polished metal. The materials are important: the bigger the refractive index difference between the 'plates' and the 'spaces', the greater the reflection at each interface. Mirrors of this kind have been produced commercially since the 1930s, by vapour-depositing alternate layers of zinc sulphide (high refractive index) and magnesium fluoride (low refractive index).

The films involved in these reflectors are so thin (about 100 nm or 0.1 µm) that they cannot be resolved by ordinary light microscopy. Electron microscopy can reveal them, but the best way to study them is to use an interference microscope. In 1967 I went to work with Eric Denton at the Marine Biological Association Laboratory in Plymouth, where Eric and Colin Nicol had worked out the way the scales of silvery fish are used in camouflage (Fig. 34). We set out to measure the thickness of the crystals of guanine that made up these mirrors. I brought to Plymouth an interference microscope that had been lent to me by Andrew Huxley (famous for his work on action potentials), who had built it himself to study the behaviour of contracting muscle filaments. At the time he was head of the physiology department at University College London and my Ph.D. supervisor. The principle of the microscope was that it directs two coherent beams of light (i.e. beams in phase with each other), one through the specimen on the microscope stage and the other through the medium adjacent to it (Fig. 51). These beams are then recombined and they interfere with each other. Depending on the difference in optical thickness (the path difference) between the specimen and medium, the interference colour of the image of the specimen changes, and this change can be cancelled out by changing the length of the optical path in the reference beam. That extra path length provides the measurement required. This sounds straightforward, but the initial coherence, and the cancellation mechanism, were both provided by the cunning use of polarizers, which I barely understood. Worse, Andrew had not told me that the elements in the high-power microscope objective were not cemented

Plate 1. Blue eyes of the Atlantic bay scallop (*Argopecten irradians*) looking out between the tentacles of the mantle. The sixty to one hundred eyes are each about 1 mm in diameter, and they cause the animal to shut if they detect objects such as fish moving nearby.

Plate 2. Female (left) and male (right) eyes of a horsefly (*Tabnanus lineola*). The facets of the upper part of the male eye are greatly enlarged to provide the improved resolution needed to detect and capture females in mid-air. The colours in the eyes are produced by chitin multilayers operating as interference reflectors. Their function is unknown.

Plate 3. Atlantic sand fiddler crab (*Uca pugilator*), on tiptoes and with its one enlarged claw raised in an attitude adopted when a female passes by.

Plate 4. Typical jumping spider showing the large principal eyes flanked by the smaller antero-lateral eyes. The (animal's) right postero-lateral eye can just be seen on the left. The lateral eyes are fixed in the carapace, but the retinas of the principal eyes move to scan the images of objects of interest. Most jumping spiders are small, 5–7 mm long.

Plate 5. The extraordinary display of the Australian jumping spider *Maratus volans*. In addition to the typical leg-waving display, males also inflate their impressively coloured abdomen so that it is visible to the female they are courting.

Plate 6. The mantis shrimp *Odontodactylus scyllarus*. The two large (3 mm) eyes are highly moveable and are seen here in different orientations. The mid-band containing the colour-vision system is visible running across the eye on the right. Below the eyes are the coloured flaps (second antennae) which act as species identifiers. Below these are the folded striking appendages.

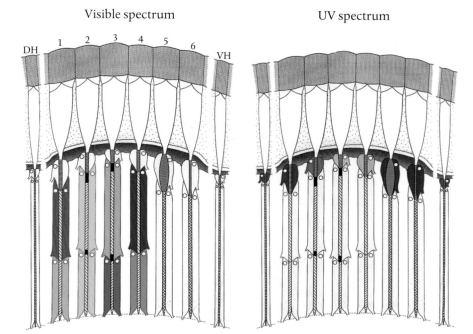

Visible spectrum   UV spectrum

Plate 7. The tiered colour-vision system in a section across the mid-band of *Odontodactylus*. The approximate colours to which the receptors respond are shown, for convenience, in the cell bodies, although the actual photopigments are confined to the rhabdoms between the cell bodies. The spectral sensitivities of the receptors spanning the visible spectrum are shown on the left in the two deeper-lying tiers. In addition to the photopigments, there are also photostable colour filters (shown in black) in the rhabdoms of rows 2 and 3. On the right are shown the four ultra-violet (UV) photopigments in the uppermost tiers of rows 1–4. Rows 5 and 6 are polarization sensitive and are not part of the colour system, although the upper tiers are also UV sensitive. The main rhabdoms in rows 5 and 6 and the dorsal and ventral hemispheres of the eye are green sensitive.

Plate 8. *The Grand Canal*, Venice, by Canaletto, about 1740.

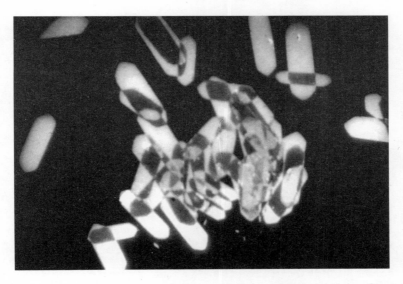

Fig. 51. Interference microscope photograph of the crystals, mainly of guanine, that make up the reflecting platelets in a fish scale (as in Fig. 52a). The dark background results from destructive interference between two light beams. The crystals introduce a path difference into one beam so that interference is no longer destructive, and they appear light. Where crystals cross, this path difference doubles, and the overlapping images become dark again.

together, and when I turned it upside down, they fell out. Panic! I did eventually manage to get them back in the right order, and I did tell Andrew about this—but not until 2 years later.

It required some more work to turn the measured path differences into actual thicknesses, and the refractive index of the crystals was needed too. This could be obtained by measuring path differences in media with different refractive indices. The upshot was that the crystals were indeed a quarter wavelength in optical thickness (Fig. 52a). We also looked at crystals from different coloured tissues. Neon Tetra fish have blue and orange regions in the iris of the eye, and the crystal thicknesses matched the wavelengths of these colours well, the 'blue' crystals being about two-thirds the thickness of those from the orange region. Squid also had quarter-wavelength plates, particularly in reflectors around the eye: these were not crystalline but made of flexible protein (Fig. 52b).

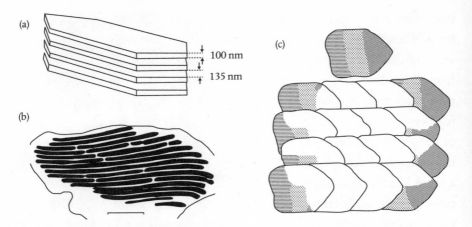

Fig. 52. a) Stack of guanine reflecting plates from a sprat scale showing the dimensions of the crystals and spaces, obtained from interference microscopy. b) Reflecting plates in the iridescent skin of a squid, from an electron micrograph. Scalebar is 1 μm. c) The overlap of scales in a herring, showing how the three colour zones in each scale overlap so that the reflected colour is white.

A clever feature of fish scales is the way whiteness is achieved (Fig. 52c). A quarter-wave reflector reflects brightly over about a third of the spectrum, so it should appear coloured—blue or green or orange. Eric had already shown that the scales of a white reflecting fish have different coloured reflectors in different parts of the scale. On the body they overlap so that at any point the scales are three-deep, with, for example, a purple-reflecting region overlying an orange region, and a blue-green region. Because interference reflectors, unlike pigments, transmit what they don't reflect, light from all three regions is reflected from the scale, and the colour of that is white. Intact fish at the fishmonger's still appear white, but when they are damaged and start to lose scales, the individual colours begin to show.

Multilayer mirrors can be tweaked in various ways to provide high reflectance or greater depth of colour. In general, quarter-wavelength stacks produce the highest reflectance over a wide spectral range. Changing the relative widths of the films and spaces to give 'non-ideal' reflectors reduces the reflectance, but it narrows the reflected waveband, and these give more saturated colours. Such reflectors are common in the wing scales of butterflies and moths, where colourful reflectors are constructed from chitin and air layers. Broader wavebands can be achieved by systematically varying the film thickness through the stack. Not all reflectors are made from thin films. The green reflector in a cat's eye is composed of layers of rods made from the zinc salt of cysteine, and the 'structural colours' in bird feathers are made from arrays of rods of melanin embedded in keratin. All of these reflectors are iridescent; that is, the colours change with angle as the path lengths of light within the stack change. In a few butterflies there is a third kind of reflector where colour does not change with angle. These reflectors are made from a three-dimensional array of spherical holes in a chitin matrix, meaning that there is no preferred direction in which interference occurs. Such structures produce bright matt colours.

I can't finish this chapter without a short tribute to Eric Denton, who made more contributions to marine science in the twentieth century than anyone else I can think of. I was lucky enough to work with him both at the Plymouth Laboratory and on the *Discovery*. He not only discovered the secrets of silvery fish but worked out how squid and cuttlefish adjust their buoyancy and how fish determine the distance and direction of sound. He discovered new visual pigments, and was the first to show that the dragonfish *Malacosteus* uses red light to illuminate prey at a depth where everything else can only see in the blue. Eric became director of the Plymouth Laboratory in 1974, a post he held until his retirement in 1987. He received many awards and was knighted in 1987 'for services to marine biology'. Eric died in 2007, aged 83. He was famous for his hospitality but also for his modesty. When asked by a journalist whether he thought his work had been important he replied: 'No, but it didn't cost much either.'

4

ESTABLISHING IDENTITY

The Problem of Recognition

Humans are astonishingly good at recognizing each other. We can recognize many thousands of faces, even if we can't always put names to them. Similarly with objects. If we study a particular class of objects, whether it is old sports cars or wild orchids, we quickly develop an ability to name them on the basis of their physical appearance. We are so good at these things that we tend to think this is an easy task, but it isn't. Just how challenging it is has become apparent since computer engineers first tried to make artificial recognition systems, half a century ago. It is true that face recognition systems at airports can check whether your face is the same as the one on the passport, but this requires that you stand in the right place facing the camera in standardized diffuse light. Then a metric analysis of features becomes possible. But these are not the circumstances we meet in ordinary life. Usually people we meet are at different distances and so have retinal images of different sizes, they present a wide variety of profiles, and the lighting conditions can vary from dim overcast to bright directional sunlight. This last complication can change the distribution of light and dark on a face in a way that almost obliterates any simple description of features, and yet our recognition system can cope with all these variations. There must be subtleties in the appearance of a face that we can still make use of, but what they are still defies computational analysis. Monkeys are like us, and can recognize faces under a variety of conditions, but to what extent this applies to other animals is much less clear.

We have massive computing power devoted to this task: tens of millions of neurons in the temporal lobe of the brain alone. For most invertebrate animals, whose brains have total neuron numbers in the range of 10^4 to 10^6, the luxury of an identification system on this scale is not a possibility. Nevertheless, many of them still need systems that can distinguish members of their own species from other species, their own sex from the opposite sex, friend from foe, edible from inedible objects, and for those that have a home, the landmarks that will enable them to find it. This chapter will explore a few of the ways that this is done. In all cases the key is simplification: restricting information to just those features of the visual world that are needed to solve a particular problem. The early ethologists, Niko Tinbergen, Konrad Lorenz, and Karl von Frisch, recognized this, and provided many examples where a specific combination of features constituted an 'innate releaser mechanism' that triggered a particular response, a 'fixed-action pattern'.[1] For example, a model bird with a long neck and short tail (a goose), flying overhead, causes no reaction in chicks, but reversing the direction of flight so that the bird has a short neck and long tail (a hawk) causes panic. Such behaviour is not learned, but is part of the genetic endowment of the animal. Not all behaviour in invertebrates and non-mammalian vertebrates is as clear-cut as this. However, the important point is that for an animal with a modest brain reducing the complexity of recognition to a few key features is crucial for making quick and effective action possible.

Fiddler Crab Solutions to Recognition

I first met fiddler crabs in 1993, on sand flats near the Duke University Marine Laboratory in Beaufort, North Carolina. There they occur in vast numbers, with each crab having a small territory containing a burrow. Male fiddler crabs have one large claw which they wave every few seconds in a characteristic arc (Plate 3). This is to attract females, who lack the enlarged claw. Most of the time the males feed—picking up wet sand with their normal claw—while waving every few seconds in a rather desultory way with the other. If a female comes through a patch of feeding males, everything changes. All of the males in the vicinity stand on their toes and wave much more vigorously, trying to entice the female to their burrow, which is never far away The urgency of this display reminded me uncomfortably of adolescence. The males are quite gentlemanly, and they do not try to coerce the females into the burrow; the decision is entirely hers. Successful pairings are rare, but when they do occur, male and female disappear down the male's burrow, where they mate and remain for at least the rest of the tidal cycle, and usually much longer.

What is it about a female that causes the males to react in this way? It is not very subtle. I pulled a small rotating cotton reel on a string through a group of feeding males, and that worked just as well as a real female (Fig. 53a). It seems that motion and size are important, but at this stage, not much else. This raises a question of why the males don't react this way to each other. It may simply be that they don't wander about in the same way as females or that males in adjacent territories know where to expect their neighbours to be or that the males' claw waving is itself an inhibitor of sexual enthusiasm in other males. As far as I know, no one has yet worked out why a female will choose a particular male to mate with. Having a large claw and waving it is a prerequisite, but given the modest eyesight of fiddler crabs, it is difficult

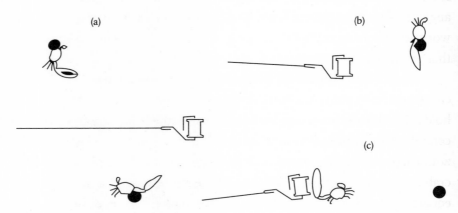

Fig. 53. Responses of males to a cotton reel pulled through their territories. a) Courtship waving by two males to the approaching cotton reel. b) Male defending his burrow when the cotton reel passes too close. c) Aggression. The crab followed the cotton reel from its burrow and is about to strike it with the back of its claw. Black dots are burrows. The cotton reel is about 2 cm long.

to know what else the females might be looking for. The males are very sensitive to intrusions into their territories, the limits of which they seem to know well, and they resist incursions by other males with threat displays and occasional fights, although these rarely cause damage. They will respond with threat to cotton reels that get too close (Fig. 53b and c), so again, they are not looking for anatomical details.

One of the most impressive behaviours of fiddler crabs is their response to predators, particularly airborne predators such as gulls. I teamed up with John Layne, then a graduate student at the marine lab, and we tried to discover what it was that distinguished potential predators from other animate objects such as conspecifics. The response to predators is very clear. The crabs first freeze; if there is further movement they run, sideways, to their burrow; and if there is still more movement, they pop into it. They will respond in this way to a human walking past, or simply rising from a crouch. We also looked at responses to vehicles on a nearby road and their behaviour to a radio-controlled toy truck, which they didn't like at all. The maximum distance at which defensive behaviour occurs gives a clue as to the acuity of the crab's eyes, and in all cases this produced a figure between 1° and 1.5°, which corresponded nicely to the minimum inter-ommatidial angle measured on the eye itself.[2] This is pretty good for a crustacean, but worse than human acuity by a factor of about sixty, so it is not so surprising that the males don't attempt to distinguish the features of passing females.

Another feature of fiddler crab eyes is that they are vertically elongated and carried on stalks above the head, about 20 mm above the ground. The crabs hold them so that the eye itself is kept vertical, and that a region around the central equator of the eye is kept accurately horizontal, in line with the horizon. This equatorial region also has the highest acuity. If you pick up a fiddler crab and tilt it through as much as 90°, the eye stays vertical, and the eye's equator horizontal. This marvellous piece of compensation seems to be controlled both by gravity receptors in the statocysts, the crab's balance organs, and by the visual appearance of the horizon itself. All this suggested to us that the distinction between potential predators and non-dangerous creatures, particularly other crabs, might have nothing to do with their appear-

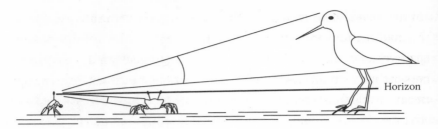

Fig. 54. Distinguishing potential predators from other crabs. Anything above the horizon must be larger than the crab and is treated as hostile. Anything below is smaller and probably another crab.

ance, but be simply concerned with whether all or part of them appeared above the horizon (Fig. 54). Lines joining the eye to the horizon define a plane at eye height, 20 mm above the sand surface. Other crabs will move around below this plane, and anything larger than a crab will intersect it. It is an extraordinarily economical way of distinguishing friend from foe.

John Layne developed an arrangement in which a fiddler crab stood in the centre of a transparent cylinder around which black squares of various sizes could be moved horizontally at different heights relative to the crab. He measured escape responses when the square moved. Although a 1° square caused little response, all larger squares (2°, 4°, and 8°) produced escape behaviour, but only when the top of the square was above the eye's horizon. Even in this artificial situation, with very unrealistic stimuli, this simple horizon rule for distinguishing predators from other crabs seemed to work. Interestingly, during waving displays, the male's claw *does* briefly appear above the horizon. This does not cause mass panic, but it might just give a passing female pause for thought.[2]

In parts of the Eastern Seaboard and the Southern Gulf Coast of the USA, several species of fiddler crab live in the same upper shore environment. They then have the problem of who to mate with: are their neighbours the right species? The physical appearances of the several species are not very different (Fig. 55 a) and, given that the crabs have fairly low-resolution eyesight, appearance alone is not going to provide the information they need. They use instead a visual signal that does not rely on acute vision: the timing of the males' wave pattern. Figure 55 shows the results of a 1968 study by Michael Salmon and Samuel Atsaides, in which they recorded the waving patterns of five *Uca* species from the same general habitat. The form of the waving pattern differs very little, certainly for the first three species. In *U. rapax* the claw is directed more sideways, and more forwards in *U. mordax*, but it isn't clear whether these differences would be detectable by a female who might not be facing the waving male. The striking differences are in the temporal patterns of the claw movement (Fig. 55b).

In *Uca pugnax* the wave is a smooth arc lasting 2 seconds. In *longisignalis*, and the other three species, it consists of a series of small staccato upwards and downwards jerks, and the overall wave lasts much longer. In *virens* the wave terminates abruptly with a single downward jerk. In *rapax* the upward

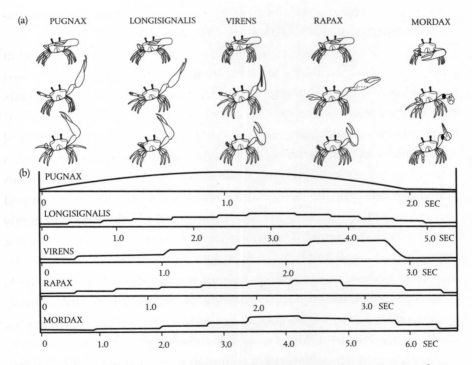

Fig. 55. Waving patterns of the males of five sympatric species of *Uca*. a) Appearance of the wave from in front. b) The timing of the different components of the waves, showing highly visible and characteristic differences.[3]

97

phase takes six jerks and the downwards phase three. The *mordax* wave is more symmetrical, rather like *longisignalis*, but the individual jerks are larger and the gaps between longer. These jerky movements are very striking, even to a human, and they provide unambiguous cues to species identity.[3]

We have seen in fiddler crabs several ways of getting round the central problem of human pattern recognition: how to distinguish objects or individuals from the geometry of their features. Males recognize females by their motion and size; they recognize potential predators by whether or not they intersect the horizon; and they recognize the waves of their own and other species less by their form than by the temporal pattern of the wave, a feature that relies little on keen eyesight but does require an ability to measure intervals of time.

Jumping Spiders: Different Eyes for Different Tasks

Apart from having eight legs, the jumping spiders (Salticidae) are different from all other spiders. They are short and stocky, catching prey by stalking rather than building webs, and they have better eyesight than any other terrestrial invertebrate. The eyes are impressive. There are four pairs: three pairs of relatively small 'secondary' eyes, and one much larger pair of forwards-pointing 'principal' eyes (Plate 4, and Fig. 58). Most, but not all, jumping spiders have dispensed with one pair of secondary eyes (the posteromedians), and the remaining four have fields of view that cover the whole space around the animal. Their main job is to detect movement. If you move, the spider will turn to look at you with its principal eyes. If you move a finger around the animal, it will follow the finger, backing off if it gets too close. This apparently inquisitive behaviour is almost primate-like. Other spiders, and indeed most arthropods, are so visually undemonstrative that the interactive nature of this behaviour comes as quite a surprise.

The main behavioural repertoire of jumping spiders had been described in the 1920s by Heinrich Homann, and in the 1950s, an ethologist, Oscar Drees, added a further twist. When a male meets a female, he does a dance

Fig. 56. Male *Phidippus johnsoni* displaying to a female.

in front of her in which he moves sideways in a series of arcs, with his front legs raised—a little like a highland fling (Fig. 56). Drees found that this courtship behaviour could be evoked with simple outlines drawn on card (Fig. 57). For a male to start a dance, the essential features were a central spot of some kind, and lines on either side signifying legs. One pair of legs would do, but more legs were better. The angle was important: about 30° to the vertical seemed to be most effective. Other shapes—cross, square, triangle, circle, or a realistic drawing of a fly—never evoked courtship, but if moved, these shapes often resulted in stalking and an attempt at capture. Evidently the legs were the thing that not only initiated courtship but also prevented attack by another jumping spider. From Drees's work it seemed that these spiders had an identification system that was reasonably straightforward. The logic seemed to be that if something moves in the field of view of the secondary eyes it is worth investigating further with the principal eyes. If it has 'legs' in the right places, it is probably another jumping spider and at the very least, should not be attacked; anything else can be pursued as potential food. Obviously, there were other categories—harmless, non-moving

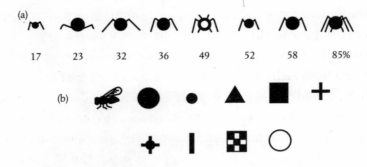

Fig. 57. a) Images used by Oscar Drees in 1952 to evoke courtship behaviour from male jumping spiders. The numbers are the percentages of trials on which the males displayed. b) Images that did not evoke courtship but which, if moved, produced capture attempts.

objects and larger moving objects that might be hostile—but the really interesting question seemed to concern the mechanism of leg detection.

When I went to Berkeley in the late 1960s on a post-doctoral fellowship, I decided to take this problem on. Jumping spiders are not exotic. In most of the northern hemisphere, the zebra spider (*Salticus scenicus*), a 7-mm-long animal with a striped abdomen, is common everywhere on walls, and especially in greenhouses. In Berkeley the easiest way to get numbers of jumping spiders was to shake the lower branches of redwoods into an umbrella. This caused some derision, but it was very effective. The commonest species in the trees was the *Metaphidippus* (now *Pelegrina*) *aeneolus*, which is small (5 mm), but larger (10 mm) animals (*Phidippus johnsoni*) could be collected from hillsides (Fig. 56). In the lab jumping spiders are very accommodating. They will last for weeks in a petri dish, if fed on fruit flies. They can be kept happy by suspending them from a strip of card waxed to the thorax, and providing them with a light card ring, which they will rotate with their feet, as though walking along a twig. This turned out to be valuable scientific tool because if you came in the door or moved around the lab, the spider would turn the ring left or right through an angle that would, if the spider had not been fixed, have brought it round to face you. This tells you straight away that the turns initiated by the secondary eyes are 'open loop' in nature. Once a particular point on the retina has been tickled by motion, the location of that point specifies a complete turn of the right size, whether or not the object that caused the turn now appears in front.

The two types of eye—the secondary and principal eyes—are fundamentally different from each other. They are all single-chambered eyes (like ours), in which the curved cornea is mainly responsible for producing the image on the retina (Fig. 58). The secondary eyes are attached to the animal's carapace and are immobile, so that the retinal receptors view fixed directions relative to the spider's body. The principal eyes, however, move. Unlike our eyes, the lenses themselves do not move, but the retinas do, so that they scan the fixed images produced by the lenses. The eye tubes are long and narrow, extending back about a third of the length of the head (strictly the cephalothorax, as there is no division between head and thorax in spiders). In juvenile spiders (Fig. 58a), the carapace is fairly transparent, and you can see the eyes moving in the head over large arcs—up to 50° in the horizontal plane. Each eye has six muscles that can move the retina up, down, and sideways, and rotate it around the visual axis (Fig. 58b). The eye movements of these young spiders are mostly conjugate, that is, they move together in the same direction, but this isn't always true; the eyes are certainly not linked mechanically.

Homann and others had already shown that painting over the secondary eyes prevented a spider from turning towards a moving object, such as a fly, unless it was already within the fields of view of the principal eyes. Covering the principal eyes, however, prevented all other aspects of vision: stalking, jumping, and courtship behaviour. The principal eyes, and their movements, evidently hold the key to the spiders' identification process. Not much could be learned from the eye-tube movements of juveniles, so I decided to make an ophthalmoscope which would let me see the retinas directly, making use of the eyes' own optics. At the same time, I wanted to be able to show the animals objects, rather as Drees had done, to see what the spiders would make of them. I also needed to know much more about the anatomy of the principal eyes, since the quality of their resolution would define what the spiders could see. I was greatly helped in these endeavours by my two fellowship sponsors. Gerald Westheimer was renowned for his knowledge of both visual optics and eye movements, both of which were very relevant here. And Dick Eakin,

(a)

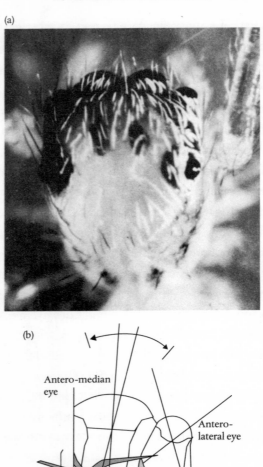

(b)

Antero-median
eye

Antero-
lateral eye

Postero-
median eye

Eye
muscles

Retina

Postero-
lateral eye

First optic
ganglion

Fig. 58. a) Head of a juvenile jumping spider, from above, showing the long eye-tubes of the principal eyes through the nearly transparent carapace. b) Diagram of the right half of the head showing the horizontal fields of view of the various eyes, and the six muscles that move the principal eyes. (Darker muscles are above the eye; lighter muscles below it.) The thick arrow at the top shows the field of view of the principal eye as it is extended by eye movements.

who I've mentioned before, was the world expert on invertebrate eyes. I could not have been in a better place. Eakin's lab team cut a complete series of sections of the head of *Metaphidippus aeneolus* from which I was able to work out the structure of the retinas and the eye muscles. This turned out to be intriguingly complicated, and I will just indicate the results here.

The principal eye retinas are unlike anything else in the animal kingdom. They are elongated vertically, so that each has a field of view about 20° high but only 1–4° wide (Fig. 59b). They are bowed slightly outwards in the body, like a boomerang, which means that the fields of view bow inwards, with the two retinae forming a vertically elongated cross (see Fig. 61). It came as a surprise to find that each retina has four layers of receptors (Fig. 59a). The deepest two layers (1 and 2) cover the whole retina, but the upper two (3 and 4) are only present in the central region. The finest resolution is in layer 1, where the receptors are less than 2 μm apart. This corresponds to an angle of 10 arc-minutes (0.17°), which beats the resolution of the best insect eyes; some dragonflies have inter-ommatidial angles of 0.25°, but in flies and bees, 1–2° is typical. In one jumping spider species (*Portia fimbriata*), which has exceptionally long eye tubes, the resolution is even better. The receptor spacing corresponds to an angle of 2.4 arc-minutes, which compares favourably with most vertebrates. The highest receptor densities are near the centre of each retina, providing a region analogous to our fovea. This region, however, is only about six receptors wide, giving a transverse field of view not much more than a degree across. It is hardly surprising that eye movements are used to extend this tiny field.[4]

What is the meaning of the four-layered retina? Two ideas have been suggested: managing chromatic aberration and estimating distance. Both seem to be right. The principal eyes, with long focal lengths, have significant chromatic aberration: short wavelengths are focussed closer to the lens and longer wavelengths further away. As early as 1975, Robert DeVoe had penetrated cells in the retina with microelectrodes, and found both ultra-violet (UV) and green receptors. More recently, work has shown that the UV receptors are found in layers 3 and 4, nearest the lens, and green receptors in layers 1 and 2. It seems that most jumping spiders

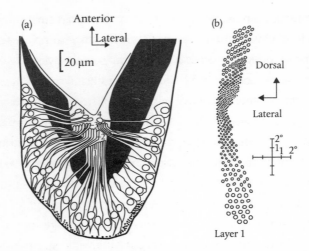

Fig. 59. Retina of the jumping spider *Metaphidippus aeneolus*. a) Horizontal section through the centre of the right retina, showing the four layers (one through four). In layers one to three, the receptive parts of the cells are the segments directed vertically on the page; in layer four, they are the oval structures. Layers three and four are sensitive to ultra-violet light; layers one and two absorb green light. Scalebar is 20 μm. b) Vertical section through retinal layer one. The receptors are closest together at the centre of the retina, where the layer is only six receptors wide.

are likely to be UV/green dichromats, with the two kinds of receptors at appropriate distances from the lens. However, a few jumping spiders are trichromats. Male spiders of the genus *Habronattus* have particularly colourful facial adornments, including red and orange, in places where only a female spider of the same species is likely to see them. A UV/green colour system would not be good at distinguishing such colours from green or yellow, so it seemed possible that *Habronattus* might have managed to extend its colour range in some way. In 2015 Daniel Zurek and his colleagues found that in *H. pyrrithrix*, a small cluster of receptors near the centre of layer one have a red filter in front of them. This changes their peak sensitivity from green to red, and provides the spider with a third colour channel, appropriate for recognizing other members of its species.[5] Such enhancement of colour vision by passive filters is known in other animals. Birds and reptiles have coloured oil droplets in their receptors, and so do mantis shrimps, as we shall see later in this chapter. One Australian genus of jumping spiders, *Maratus*, is not only colourful, but the males extend flaps on the abdomen during courtship to create an astonishing visual display (Plate 5). Whether *Maratus* has a retinal filter system like *Habronattus* is not yet known.

Another strange feature of the principal retinas is the presence of two complete receptor layers, 1 and 2, both responsive to green light. What can two sets of receptors provide that one cannot? A very plausible answer, provided by Takashi Nagata and his colleagues in Japan, is that the two layers act as a focus detector for determining distance. The idea is that for distances between 1 and 3 cm, the range over which spiders jump, the image on layer 1 remains in focus within the length of the receptors, but the image on layer 2 becomes progressively more out of focus as objects get nearer. This means that the difference in contrast for edges between the two layers changes with distance in a way the spiders should be able to measure. Nagata showed that jumping with just one principal eye open remains accurate, so stereopsis is not involved, and furthermore, changing from green to red light degrades jumping performance in a way predicted by the fact that the focus for red light lies deeper in the retina than for green light.[6] This type of distance-from-focus mechanism is known in some other animals, most famously in chameleons, although there the focussing is done by an active change in the lens. In jumping spiders the focus of the lens is fixed, but image position varies with distance.

We now reach the most interesting question: how do jumping spiders use their narrow retinas to find out whether the object they are facing has the key features of another jumping spider? We know from Drees's work that this has something to do with legs, but how are these detected? These are not straightforward two-dimensional retinas, where one could imagine the fitting of some innate internal template of a figure with legs to the appearance of the seen object, rather like the face-recognition cameras at airport security. The principal retinas move, and this motion is likely to have a bearing on the recognition process. It was with this in mind that I set out, in

1968, to devise a way of visualizing the movements of the principal eye retinas, while they were looking at objects of interest. It would be an understatement to say that I was amazed by what I found.

Designing and making an ophthalmoscope to look into spiders' eyes was quite a challenge (Fig. 60). As with a medical ophthalmoscope, I used the spider's own lens to form an image of its retina. This image is at some distance outside the eye, and it had to be brought to a focus by another lens, and this image viewed with an eyepiece, as in a telescope or microscope. At the same time, the eye had to be illuminated, partly so that I could see the retina, but more importantly so that I could show the spider objects that I hoped it would react to (Target, Fig. 60), against a reasonably wide uniform background. This meant that the instrument had separate viewing and illuminating channels, combined using a half-silvered beam-splitter. Once

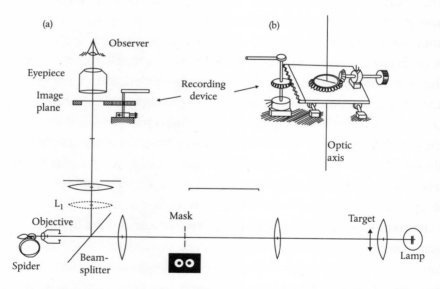

Fig. 60. Ophthalmoscope for viewing the spider's retinas. a) Diagram showing the viewing and illuminating light paths. The illuminating beam projects an image of the target onto the spider's retina; the mask prevents reflections from the corneas. The beam-splitter allows the viewer to see the retina, while the spider sees the target in the other beam. In the viewing beam lens L_1 can be inserted to view the eye surface. b) Arrangement for recording rotational and horizontal retinal movements. The record here is unusually long. Most scanning bouts last 2–5 seconds. Scalebar is 10 cm.

I had got this more or less working, I still couldn't see much, because there were reflexions from the spider's own lens and various other lenses in the system, all of which veiled the image I was trying to see. The trick turned out to be to put black spots on microscope slides at strategic places in the light path, where they intercepted the images causing the reflections, without blocking too much of the viewing beam itself (Mask, Fig. 60). When all that was done, I started to see things.

Looking into the secondary eyes—the postero-laterals and antero-laterals (Fig. 58)—produced the kind of images I had expected. It was possible to make out the outlines of the receptors in a regular hexagonal array. The spacing was about 1° in the postero-lateral eyes, but the antero-lateral eyes had a higher density region in the forwards-pointing part, where the receptor spacing came down to 0.6°. The principal eyes were quite different. There was no trace of the regular mosaic that retinal layers 1 and 2 might have produced. Instead there was a feathery pattern of iridescent bluish lines, making a fan-shaped pattern (Fig. 61a). Comparing this to the sections of the eye, the only structures that made the same pattern were the parts of the layer-4 receptors that protruded out into the eye tube (seen at right angles to the section in Fig. 59a). The layer-4 receptors are confined to the central part of the retina, but it was possible to reconstruct the outline of the rest of the retina from the histological sections, and these outlines are shown in Figure 61a, superimposed on a photograph of the pattern of the layer-4 receptors. It was not easy to get this photograph, because the images were dim and the retinas usually in motion.

It was an exhilarating but very weird experience to look into the moving eyes of another sentient creature, particularly one so far removed in its evolution from oneself. As I had hoped, when I introduced a black spot into the field of view the first thing the eyes did was to centre the two retinas onto it in a fast movement resembling the kind of saccadic eye movement we make when fixating a new object. If the dot moved, they tracked it. What came next was the real surprise. The two retinas scanned the dot by moving repeatedly side to side across it, with a regular period of about 2 seconds. At the same time, the two retinas rotated together, through an angle of about

50°, again in a repeated motion with a longer period of between 5 and 10 seconds (Fig. 61b and c). This scanning pattern usually lasted for a few seconds, before the eyes broke away and looked elsewhere; the record in Figure 61c is unusually long. The pattern was remarkably consistent from one presentation of the spot to the next, and presenting different shaped targets did not seem to change it, except that with wider stimuli the horizontal component of the scanning lengthened, so that the retinas appeared to be scanning from one edge of the target to another.[7]

In the pre-video days of 1968, I did not have a movie camera sensitive enough to capture the retinal movements and I had to resort to using myself as the recording instrument. I devised an arrangement consisting of a rotatable line on a Perspex disc, which was mounted on a rack and pinion so that the disc could also be moved laterally with a lever (Fig. 60b). Both rotational and sideways movements were recorded by potentiometers, and the signals fed to a chart recorder. I could follow the side-to-side movements with one hand,

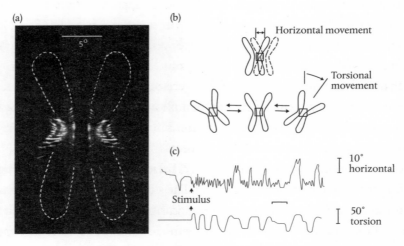

Fig. 61. a) Ophthalmoscope view of the principal eye retinas. The visible structures are the receptors in layer four (see Fig. 59a). Layer one outlines have been superimposed. In fact, only the right retina was photographed and its mirror image added to provide the combined field. b) Diagram of the two types of scanning movements seen when a target is presented. c) Recordings of the retinal movements made with the device illustrated in Fig. 60. Scale bar is 10 s.

and the rotations with the other. I can't pretend that these records are perfect, but I'm happy that they are a reasonably accurate record of what I saw.

What was the spider trying to do? It struck me at once that the eyes were carrying out some kind of program for extracting relevant features from the target. From Drees's study (Fig. 57), it was clear that one of these features, though probably not the only one, was the presence of a pattern of contours that could be construed as legs. If we suppose that the rows of receptors in the retinas are wired up in the brain to form vertical line or edge detectors, similar to those in the primate visual cortex, then the retinal movements become comprehensible. To detect an edge, the detectors must move across the figure (the side-to-side movements), and to detect its orientation they must also rotate. I assume that having repeated this scanning pattern for a few cycles the spider will have accumulated enough information from the line detectors to allow it to come to a conclusion: either there is enough evidence for 'legs', or there isn't. The test would be to show a male some of Drees's drawings while watching both its retinal movements and its behaviour, but I never came up with an arrangement that would allow me to do this.

This scanning system does seem to make sense of some other aspects of jumping spider behaviour. The males' courtship dances (Fig. 56) involve both lateral motion, and a vertical exaggeration of the leg pattern, as though the males were trying to amplify the kinds of signals that the scanning system is designed to detect. Females, it seems, are always hungry, and the best protection a male has is to make absolutely sure that he is recognized as another jumping spider. It may also explain why some jumping spiders go in for what appears to be self-mimicry. The striped pattern on the abdomen of the common zebra spider (*Salticus scenicus*) and the patterns on the face of the Californian *Metaphidippus manni* seem to function as extra legs, again emphasizing the pattern of contours that the scanning system should pick up (Fig. 62). Colour is certainly important too, particularly in *Habronattus*, mentioned earlier, and the Australian peacock spiders (*Maratus*), whose blue, red, and orange abdominal gyrations have made them YouTube celebrities (Plate 5).

Fig. 62. Striped markings on the bodies of jumping spiders. a) The abdomen of *Salticus scenicus*, and b) the face of *Metaphidippus manni*. These can be thought of as imitating and augmenting the normal pattern of the legs. Scalebar on (a) is 1 mm.

Most jumping spiders eat insects, but there is one genus of tropical rain forest spider, *Portia*, which specializes in hunting other spiders. These remarkable spiders have been studied extensively by Robert Jackson and his colleagues in New Zealand, who emphasize the extreme flexibility of their behaviour. *Portia* tailor their hunting strategy to the particular kind of spider they are trying to catch. They detect other jumping spiders by the smell of the silk 'draglines' that these spiders habitually trail behind them. They then locate their prey visually, which they recognize not only by the legs but by the presence of at least one round eye. Thereafter they stalk their prey with a particularly strange but cryptic walk, until close enough to strike. For web-building spiders, they use a variety of tactics which involve imitating the vibration patterns of a struggling insect, the aim of which is to lure the spider to the edge of the web, where it can be attacked. Often this behaviour involves the use of detours. These require routes in which the prey is not continuously visible, and so must involve a degree of understanding of the local environmental geography. When attacking an orb-web spider such as *Argiope*, these detours can last as long as 20 minutes. Jackson has pointed out that if *Portia* were a vertebrate, we would be using words like *intelligence* and *cognition*, rather than the language of stereotyped responses usually reserved for arthropod behaviour.[8]

The use of a narrow scanning detector to interrogate the features of objects of interest is very different from the way that most visual systems work. It is not unique to jumping spiders, but it is rare. Something similar occurs in one genus of copepod crustacean (*Labidocera*), some predatory sea snails, and a water beetle larva (*Thermonectus*). The animals we come to next, the mantis shrimps, have a hybrid visual system in which seemingly ordinary apposition compound eyes also incorporate a scanning array. These are probably the strangest eyes in the animal kingdom.

The Hybrid Eyes of Mantis Shrimps

Mantis shrimps, or stomatopods, are crayfish-sized crustaceans that live mainly in tropical coral reefs (Plate 6). They are malacostracans, the same class as the prawns and lobsters, but they have had a separate evolutionary history for the last 400 million years. They are impressive predators, making use of a unique weapon, the second thoracic appendage, which can be used as either a spear or a club (Fig. 63). The final segment of this appendage (the finger) is either spiked or knife-like, and is normally folded back into a

Fig. 63. A mantis shrimp (*Squilla*) showing striking appendages on the second thoracic segment.

groove in the penultimate segment. When attacking a soft-bodied animal such as a shrimp or fish, this finger is rapidly extended to spear the prey, and then equally rapidly returned to its serrated groove. When dealing with hard prey, such as a crab or a snail, the finger is kept sheathed, and the armoured joint between the penultimate segment and the finger is used as a club. This again can be extended at enormous speed, to smash the shell of the prey. Some of the larger stomatopods are credited with being able to break aquarium glass with such a strike. Different species tend to specialize as either spearers or smashers.

In common with other predators, such as jumping spiders, stomatopods need ways of recognizing potential mates and avoiding too much conflict with other members of their own species (conspecifics), especially those living in nearby burrows in the coral. According to Roy Caldwell and Hugh Dingle, who have studied stomatopod behaviour in both the wild and the lab, mantis shrimps only recognize each other but also have quite elaborate rituals for establishing local hierarchies, and thereby coming to arrangements that minimize further territorial fighting. When two stomatopods first meet, however, there are fights. These involve each individual hitting the other with their clubs, on the telson, the heavily armoured last abdominal segment. These are rarely damaging, but they do establish who is strongest, and who needs to give way. Courtship seems to be partly a matter of the female *not* provoking a fight, but coming up to the male and digging her head under his body. If this is successful, the two will pair and live in the same burrow, where they will mate and may maintain a brood in the burrow until the hatchlings are ready to become planktonic larvae. Mantis shrimp defend their burrows fiercely, either by presenting a front view with all their weaponry displayed or by blocking the entrance with their armoured telson. They often block the burrow entrance with debris at night. Many stomatopods are highly coloured, which no doubt helps in species recognition. In particular some have large 'meral spots' on the basal segment of the striking appendage, and these have different colours depending on the species. The flap-like second antennae are usually brightly coloured (Plate 6), as are the uropods next to the telson on the last abdominal segment. As we shall see, some of these structures also polarize light, in patterns that the eyes can detect.

The eyes of mantis shrimps are unique. They are quite large, either round or oval, with a conspicuous stripe—the mid-band—through the centre (Fig. 64). The first thing one notices is that they are rarely still, often making large, fast movements in which the axis of the eye moves horizontally, vertically or obliquely. There are also movements in which the eye rotates around its axis (Fig. 64). The strangest thing for an observer, used to human eyes that move together, is the extreme independence of the two eyes: they rarely move in the same direction or at the same time. The exceptions to this are when both eyes are tracking the same object or when the animal is preparing a strike. There is another class of movement, referred to as 'scanning'. These are small, slow movements, made at right angles to the mid-band. They are the means by which the mid-band, which only has a vertical field of view of a few degrees, samples an object for colour and polarization information. The activity of the eyes, particularly when something new appears in the field of view, gives the impression of almost primate-like inquisitiveness.[9] To make sense of all this, we must now turn to the optical organization of the eyes themselves.

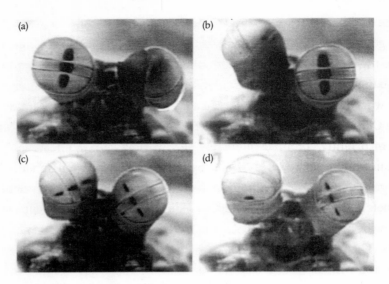

Fig. 64. Eye movements of *Odontodactylus scyllarus*: stills from about 30 seconds of video. Notice the independence of the eyes, the changes in the eyes' direction and rotation around their axes, especially between a) and c). In a) and b), where one eye is looking directly at the camera, large triple 'acute zone' pseudopupils are visible. The diameter of each eye is about 5 mm.

Each eye of a mantis shrimp is really a combination of two very different kinds of eye—the two hemispheres and the mid-band—each with a quite different function. The upper and lower hemispheres are more or less conventional apposition eyes. They have inter-ommatidial angles between 1° and 5°, and contain two receptors types, which, as in many other crustaceans, respond best to green and UV light. Both hemispheres have an 'acute zone' where the inter-ommatidial angles are smaller, about 0.5°, and which they direct at potential prey before a strike. The increased ommatidial density in this region gives rise to an enlarged dark pseudo-pupil in each hemisphere (Fig. 64a and b). The presence of the two acute zones, separated by several millimetres, means that each eye should be capable of estimating the distance of a target, in much the same way that we use the differences between the images in our two eyes to estimate depth. Sigmund Exner, studying the European stomatopod *Squilla mantis*, had reached this conclusion in 1891: mantis shrimps have binocular vision in a single eye.

The mid-band could not be more different. It consists of just six rows of enlarged ommatidia running across the whole width of the eye. All six rows share the same narrow vertical field of view. The first indication that the mid-band might be doing something special came when Justin Marshall, then working with me in Sussex, made sections of the mid-band of an eye from *Odontodactylus scyllarus*. He found that the ommatidia contained cylinders of brightly coloured pigment—yellow, orange, red, and blue. These were not photopigments (i.e. rhodopsins), but stable pigments that did not change in bright light. We thought they must be filters, like the coloured oil droplets in the cones of birds' eyes. But if that were true, it would imply that they were filtering the light before it reached the photopigments themselves. The obvious next step was to find out what these photopigments were, and why they needed filters in front of them. The best instrument to determine what photopigments are present in living receptors is a micro-spectrophotometer. These machines are not common, not cheap, and are far from easy to use. We had no such facility at Sussex, so in 1987, Justin went to Baltimore for a few months to work with Tom Cronin, who had such an instrument, much experience, and was interested in stomatopods. The micro-spectrophotometer itself consists of a microscope through which a narrow beam of light of variable wavelength is focussed onto a receptor on a microscope slide. Beneath the receptor is a photocell to measure the light passing through the receptor. The photopigment in the receptor absorbs light from the part of the spectral range to which it is sensitive, so by measuring what light is left after passing through the receptor the absorption spectrum of the receptor's photopigment can be estimated. This sounds straightforward, but the receptors are very narrow (a few micrometres), and the procedure must be carried out in near darkness to avoid bleaching the photopigment before the measurements are made. I have no experience of doing this myself, but I greatly admire those who have the skill.

The story that Justin brought back from America was amazing. These eyes contained eight different visual pigments in cells of rows 1–4 of the mid-band (Plate 7). To put this in perspective, our colour vision system is based on three visual pigments; some birds and butterflies may have as

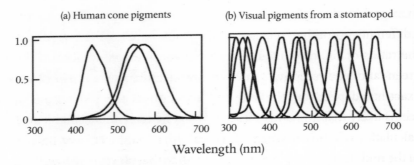

(a) Human cone pigments   (b) Visual pigments from a stomatopod

Wavelength (nm)

Fig. 65. a) The sensitivities to different wavelengths of the three cone pigments in man. b) The sensitivities of the twelve visual pigments in a mantis shrimp (Neogonodactylus oerstedii).

many as five, but *eight*! This was not even the total number. A decade later, using microelectrode recordings from single receptors, Justin and Johannes Oberwinkler found four more receptor types, and this is time sensitive in the UV range (wavelengths between 300 and 400 nm). This brought the total to twelve (Fig. 65). This was a colour vision system quite unlike ours, or any other known animal.[10]

Plate 7 shows the way the different colour receptors are organized in the mid-band. There are three tiers of receptors in each row. In the lowest tier, the rhabdom—the central narrow cylinder bearing the photopigment—contains contributions from four receptors, and in the middle tier from three receptors (except row 2, where, for some bizarre reason, the numbers are reversed). The small topmost tier has a single receptor, and its rhabdom contains an ultra-violet-sensitive pigment.[10] So each tier of each row contains a different photopigment, making twelve in all. In Plate 7 the receptor cells have been coloured, very roughly, in the colours to which their rhabdoms are sensitive. (Strictly speaking, the colours relate to the photopigment in the narrow rhabdoms *not* the receptor cells themselves, as they are illustrated in the figure.) The four photostable filters, which first attracted Justin's attention, are in rows 2 and 3, between each tier. They are black in the figure, but they are actually brightly coloured.

Why should these creatures have a twelve-visual-pigment colour vision system when we ourselves, most vertebrates, bees, and many other animals manage with three pigments? It seems that the mantis shrimps simply have a different approach to seeing colour. In our system colour is determined by the ratios of the responses of the three receptor types. Stimulation both green- and blue-receptive cones will produce a blue-green sensation whose exact hue depends on the ratio of the responses of the two cone types. It seems likely that in mantis shrimps, the system is simpler, and it avoids the calculation of ratios. If, when presented with a coloured object, the receptors of the middle tier of row 2 respond best, then that object is yellow. If the maximum response is from the lower tier of row 4, then the object is deep blue. With that many receptor types, no comparisons are required.[11]

A system like this works best if the spectral ranges over which the different photopigments respond are quite narrow, so that each colour is registered distinctly from the others (Fig. 65). Human cone pigments each respond to about a third of the visible spectrum (roughly 100 nm), and although that is ideal for a system based on ratios, it won't do if the aim is to identify a hue from the response of a particular receptor. In fact, the responses of stomatopod photoreceptors *do* have unusually narrow wavelength ranges. These narrow spectral ranges are the result of filtering, which is where the photostable filters in rows 2 and 3 come in. They cut out short-wavelength light, pushing the responses of the photopigments beneath them towards the red end of the spectrum, and at the same time, narrowing the spectral ranges over which they respond. For receptors in the lowest tier, the photopigments in the middle tier also acts as filters, with much the same effect as the photostable filters. These various filter effects tweak the performances of the different photopigments so that they respond over narrow wavelength ranges.

The colour vision of mantis shrimps has been measured, by training them to attack different coloured objects, and it is rather poor. In terms of the smallest differences in hue that can be detected, it is much worse than in either man or bee. This fits with the idea that stomatopods simply identify colours by which receptors are most stimulated. On such a scheme, the number of identifiable colours would correspond to the number of receptor types—just eight in the visible spectrum—and that fits the training results quite well. If stomatopods were like us, and took ratios between the outputs of the different receptors to determine the colours of objects, then we would expect their colour vision to be exceptionally good because of the numbers of receptors involved. It is a relief to find that it isn't.

We have not finished with the mid-band; there are still rows 5 and 6. They are not concerned with colour, but with the analysis of polarized light. This is a faculty that we do not possess, but it is common in insects and crustaceans. They use it mainly for navigation. The sky contains a pattern of polarized light closely related to the position of the sun, and if the sun is not visible, this pattern can be used in its place. Mantis shrimp probably do not use polarization for navigation, but they use it as an extra cue, a bit like colour, for detecting patterns on the bodies of other mantis shrimps.

Polarized light comes in two forms: linear and circular. Many animals can make use of linearly polarized light, but the remarkable, and probably unique, feature of rows 5 and 6 of the stomatopod mid-band is that they can deal with circular polarization as well. This discovery was made by Tsyr-Huei Chiou and his colleagues in 2008.[12] Their paper begins dramatically: 'We describe the addition of a fourth visual modality in the animal kingdom...'.Unfortunately, the physics of polarization is not especially easy, but to do justice to this remarkable capability, I feel I should try to provide at least a minimal explanation, beginning with linear polarization.

Justin Marshall realized as soon as he looked at sections of rows 5 and 6 with the electron microscope that their structure had something to do with polarization. The rhabdoms had very well-ordered microvilli—the thin rod-like processes whose membranes bear the photopigment rhodopsin. The microvilli are in blocks in which they are all parallel, but from one block to the next, the orientation of the microvilli changes by 90° (Fig. 66c). Such an arrangement is found in the eyes of bees and other arthropods that are known to make use of the polarization patterns in skylight. To understand the significance of the arrays of microvilli, we first need to look at the way photons interact with the photopigment molecules. The energy of a single photon consists of an electrical and a magnetic field vibrating at right angles to each

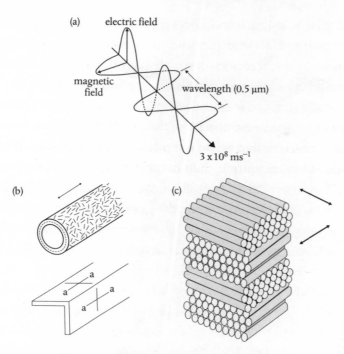

Fig. 66. a) Diagram of a photon, showing electric and magnetic fields. The direction of the electric field is known as the E-vector. b) In a microvillus from a rhabdom, the majority of photopigment molecules are aligned with the long axis. This is better shown in a square section (below), where two axes (a–a) are aligned along the structure, compared with only one axis in each of the other two dimensions. c) Part of a rhabdom showing blocks of microvilli at right angles to each other.

other (Fig. 66a). Conventionally it is the direction of the electrical component of the wave, known as its *E-vector*, which specifies the photon's polarization direction. A rhodopsin molecule is most likely to absorb a photon if the excitable double-bond in the molecule is aligned with the E-vector of the photon. It is a feature of a cylindrical membrane, such as a microvillus, that the majority of the rhodopsin molecules embedded in it will have their double bonds aligned parallel to the long axis of the cylinder (Fig. 66b). This means that a rhabdom with parallel microvilli will be more sensitive to photons with e-vectors aligned with the microvilli in the array. In turn the receptor that bears the array of microvilli will be most sensitive to light polarized in the plane containing the long axes of the microvilli, and less sensitive to light polarized at right angles. If receptors are paired, so that the microvilli in one are at right angles to those in the other, then the direction of polarization in the light reaching the receptors can be obtained by comparing the outputs of the two members of the pair. This is the way bees extract the polarization direction in skylight for use in navigation. It is also the way that the receptors in the upper and lower hemispheres of mantis shrimp eyes are arranged.

The main rhabdoms (receptors 1–7) in rows 5 and 6 are also arranged in this way, in paired blocks that can function as analysers of linearly polarized light (Fig. 66c). But this arrangement will not work for circularly polarized light. This is because in circularly polarized light the e-vector circles through all directions (Fig. 67c), and so for a block of oriented microvilli it might as well be unpolarized. The way that circularly polarized light can be

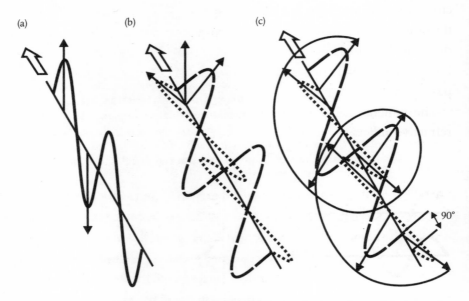

Fig. 67. Linear and circularly polarized light. a) Linearly polarized light with a single E-vector. b) Two linearly polarized waves in phase with each other will add to give a wave with a single E-vector as in a). c) If two linearly polarized waves are a quarter-wavelength (90°) out of phase, the E-vector of the combination rotates around the direction of travel, as can be seen from the way the maximum of one wave coincides with the minimum of the other. This gives rise to a circularly polarized wave; when seen end on, the E-vector of the wave describes a circle.

made detectable is to convert it into linearly polarized light. This is what happens in rows 5 and 6, as a result of the remarkable properties of the small eighth rhabdoms lying above the main rhabdoms in the light path. To explain this we have to look into the way circular polarization comes about. The next three paragraphs are not for the faint-hearted.

Imagine two light waves, linearly polarized at 90° to each other, travelling along the same path. If the waves are in phase with each other, they will add to give a single wave linearly polarized at 45° to the original waves (Fig. 67a and b). Now suppose that the two waves are not in phase, but one lags the other by a quarter of a wavelength. They will still add, but because the two waves are never in step, there can be no single e-vector in the combined wave. Instead, the e-vector rotates around the direction of travel of the wave, which, when seen end-on, describes a circle, as shown in Figure 67c. This is circularly polarized light, and depending on whether one wave leads or lags the other by a quarter-wavelength, the spiral generated will be 'left-handed' or 'right-handed'. How can this circularly polarized light be converted to linear polarized light that arthropod rhabdoms can detect? The answer is to pass it through an optical device known as a quarter-wave plate.

The mineral calcite has the interesting property that it has different refractive indices for light polarized in different planes. This is why you see

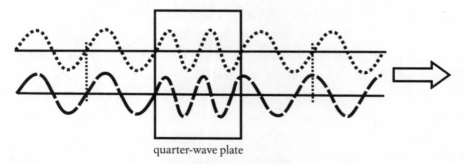

quarter-wave plate

Fig. 68. Quarter-wave plate. If two in-phase waves polarized at right angles pass through a calcite plate of the right thickness, one will be slowed relative to the other by a quarter-wavelength. When recombined they will be circularly polarized, as in Figure 67c. Passing a circularly polarized wave in the opposite direction (right to left) will result in a linearly polarized wave.

two images if you look through a calcite crystal at print on a page. The difference in refractive index means that the crystal slows down light polarized in one plane more than in the other. A calcite crystal can be cut to a thickness in which the difference in speed of the two polarized waves results in a difference of a quarter of a wavelength between the waves when they emerge (Fig. 68). As we have seen, a quarter of a wavelength is the difference in phase between the two waves that make up circularly polarized light (Fig. 67c). Conversely, the effect of putting a calcite quarter-wave plate into a beam of circularly polarized light is to cancel out that phase difference, which turns the beam into linearly polarized light. Such a beam could then be detected by biological analysers made of microvilli (Fig. 66c). Chiou and his team discovered that densely packed microvilli have a similar property to calcite, namely that the refractive index is different for light polarized in different planes. This arises from the difference in the packing of the membranes in planes parallel to and across the axes of the microvilli. In other words, structures such as the blocks of microvilli in the eighth rhabdoms of rows 5 and 6 could behave as quarter-wave plates, provided the blocks were of the right thickness. The eighth rhabdoms in rows 3 and 6 are both about 100 μm thick, and measurements made by polarization microscopy showed this to be the right thickness to introduce a quarter-wavelength difference between the wave pairs making up circular polarized light.

The plane polarized light that emerges from the eighth rhabdom will have a different E-vector direction depending on whether the rotation direction of the entering circularly polarized light is right-handed or left-handed (Fig. 69b). This difference can then be detected by the rhabdoms below, in the underlying receptors 1–7. The fact that the axes of the microvilli in the eighth receptors of rows 5 and 6 are at right angles to each other means that between them, the receptors in the two rows can distinguish unambiguously the 'handedness' of the light they receive. Chiou's team went on to successfully train mantis shrimps to distinguish right- and left-handed circularly polarized light. Finally, in this amazing *tour de force*, they showed that patterns on the telson and uropods—the flaps on the last abdominal segment—of mantis shrimps (*Odontodactylus cultrifer*) differentially

(a)

R8

R1–7

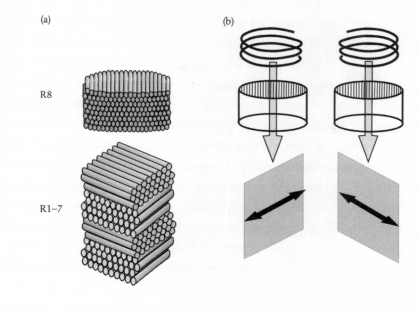

(c)

L

R

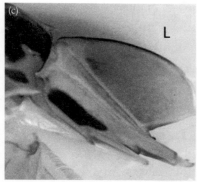

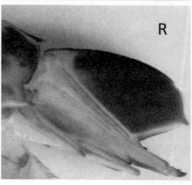

Fig. 69. a) Diagram showing the directions of the microvilli in the receptors of row 6. Row 5 is similar except that the microvilli of R8 are at right angles to those in row 6. b) The block of microvilli of R8 in row 6 acts as quarter-wave plate, which converts left- and right-hand polarized light into linearly polarized light with E-vectors at right angles to each other. c) Part of the telson of *Odontodactylus cultrifer* photographed through left- and right-hand circularly polarizing filters.[12]

reflect left- and right-handed circularly polarized light (Fig. 69c). Interestingly, these patterns are found in males but not females, suggesting that they are involved in sexual signalling.[12]

To sum up: the mid-band of stomatopods contains a battery of sensors capable of dissecting the spectrum of light reflected from other animals or objects, in both the visible and UV. In addition to this, detectors in rows 5 and 6 can analyze patterns of polarization, both linear and circular. The detection of circular polarized light is unknown elsewhere in the animal kingdom, so this submodality can be thought of as 'private' to stomatopods. A problem with this impressive array is that it only images a single strip of space a few degrees wide. The mid-band does not have a two-dimensional image of the surroundings, like that provided by the upper and lower hemispheres of the eye. It follows that the mid-band sensors can only be useful if they can cover, by scanning, an area big enough to provide identity cues from an object such as a prey animal or another stomatopod. The eye movements made by O. scyllarus are of many kinds. They include fast movements that target new objects, movements to track objects, and optomotor responses to movements of the whole surround. These eye movements are common to many animals, but there was a fourth kind that I had not seen before. These were repeated small movements about 7° in extent, and quite slow ($40°s^{-1}$). They could be in any direction, but were mainly at right angles to the mid-band, and they often involved rotations of the eye around its axis. These are almost certainly the scanning movements that enable the mid-band to extract colour and polarization information.

When we look at a mantis shrimp, such as that shown in Plate 6, what we see is a creature with eyes at the top, colourful second antennae, narrow first antennae, and striking appendages below, all in a particular geometrical layout. To us the spatial layout of the parts of the head is of primary importance in identifying the animal as a mantis shrimp, but to another mantis shrimp, this layout may not even be registered. We may recall that male fiddler crabs cannot distinguish a female from a cotton reel, and there is no reason to suppose that mantis shrimps are any better at detecting subtleties of geometry.

We can be certain, however, that they will pick up particular badges of identity, notably colours and patterns of polarization, that will reliably establish whether what they are looking at is a conspecific (to be treated with caution) or something else (perhaps to be eaten).

One feature that links mantis shrimps, jumping spiders, and humans is a division of labour between different parts of the visual system. In ourselves and mantis shrimps, the bulk of the eye is concerned with detecting salient objects, especially those that move, and then redirecting the eye so that the object falls onto a special region—the fovea in our eyes or the mid-band in stomatopods—which then performs a more detailed analysis. In jumping spiders the division of labour is more obvious. The secondary eyes are essentially movement detectors which redirect the head to face the source of movement. The principal eyes then scan the object for the presence of key contours, and in some cases colour. I find the jumping spider solution—having different eyes for different jobs—particularly neat because it saves a great deal of space. The low-resolution task of detecting motion is relegated to small eyes with limited resolution, while the feature analysis is dealt with in long 'telephoto' eyes with narrow fields of view and high resolution. In some ways the mantis shrimp solution is the least satisfactory, because the eyes have to 'time share' between the different tasks of picking up external movement, which needs a stationary eye, and scrutinizing objects with the mid-band, which requires scanning movements. It seems unlikely that one eye can do both things at the same time. Perhaps having two eyes that can move independently (Fig. 64) may be an advantage here.

The three animals I have discussed in this chapter all have ways of identifying different classes of object, but these do not seem to depend on recognizing the spatial layout of particular features, which is a computationally expensive process. Instead they use simple cues that provide enough information to place other animate objects into biologically important categories—conspecifics, prey, and predators. Motion relative to the background is of prime importance, as motion is nearly always of biological origin. Thereafter specific cues are used. Fiddler crabs use position relative to the horizon to detect predators, and the timing of claw waves to determine who to mate with; jumping spiders distinguish conspecifics from prey by the pattern of legs, and in some cases, colour, using a scanning system; and mantis shrimps make similar judgements using colour and a very sophisticated system of polarization analysis. These 'simple' systems show what can be done without the computational power of a large brain.

5

WHERE DO PEOPLE LOOK?

While working on the eyes and visual capabilities of different animals, I was often drawn to make comparisons between their eyesight and my own. I became aware that there was much we didn't know about human vision, and in particular the way we use vision in everyday life. Much of the study of human visual perception in the nineteenth and twentieth centuries had taken place in rather controlled laboratory conditions, to the detriment of more naturalistic observation. Being more of a naturalist than a psychologist, my inclination was to observe behaviour rather than to manipulate it. It was in the field of what one might call visual natural history that I decided, from about 1990, to devote more of my efforts. Of particular interest to me was the way we make use of eye movements to get the information we need to guide and control our actions. Prior to 1990 the only field in which there had been a systematic study of eye movements was reading. As early as 1920, it had been shown that we take in written text not by moving our gaze smoothly along the line, but in a stop-go manner, where about seven letters are taken in during each stationary fixation, after which a fast jump (known as a saccade) moves the centre of gaze further along the line. However, apart from reading, there had not been more than a handful of studies that addressed the way the eyes are used in everyday tasks. This is in contrast to a great deal of work on how the eye movement *system* works. The structure of the eye muscles, the dynamics of the eyeball, and the anatomy and neurophysiology of the brain structures that control eye movements were all known about in impressive detail.[1]

The nice thing about working on humans, compared with other animals, is that they can tell you what they see. Rather than simply relying on what one can infer from an objective knowledge of the visual system, or what laborious training can establish, humans can actually tell you what the world looks like. These subjective and objective views can be very different. When we look around a room or at a landscape, our impression is that we see a fully detailed scene, across which we let our gaze wander smoothly, like a panning shot in a movie. Although this is indeed what the mind's eye sees, it is not the same as the image the retina provides. Fully detailed resolution is confined to the fovea, which is only about 1° across—the width of a fingernail at arm's length. Beyond the fovea resolution falls away dramatically (Fig. 70). The detail we think we see is the result of eye movements that direct the fovea around the scene, so that we always see detail in the direction of our current line of sight.

The smoothness of the way we look around is also an illusion. Our eyes move in discrete jumps (known as *saccades* from the French word for *jerk*). The image the retina receives is discontinuous, and provides a series of snapshots that our brain somehow joins up to form a unified view. What we 'see' is a heavily edited version of the retinal image.

Fig. 70. View of a room scene which has been blurred in a way that mimics the fall-off in resolution with increasing distance from the fovea. In any single fixation, only the image in the central 1–2° around the fovea is clearly resolved. This picture represents a field about 40° wide.

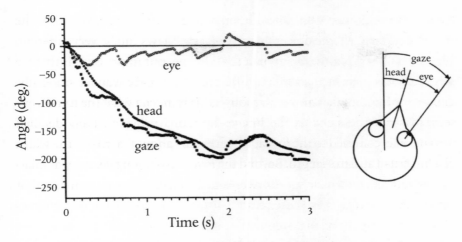

Fig. 71. Record of the eye, head, and gaze movements made by the author while looking around a room. Most of the gaze change is made by the movements of the head. The eye makes fast saccades at a rate of about three per second; between saccades, the eyes make movements that are a mirror image of the head movements. These eye movements add to the head movement to produce gaze changes, which contain saccades and stabilized 'fixations' in which the gaze angle hardly changes. It is during these 'fixations' that visual information is acquired.

The jumpy nature of the image can be seen in Figure 71, which is a record of the movements of my left eye and head as I looked round a room. The third record—gaze—is the direction of the line of sight of my fovea in space, and it is the sum of the head and eye movements. The record shows that the gaze changes direction in saccadic steps and that between these saccades, gaze is almost stationary. These stationary periods are referred to as fixations, and it is during these fixations that we take in visual information. It is clear from the record in Figure 71 that the large sweeps of gaze are mainly the result of head movements. The eyes do two things: they are responsible for making the fast saccades, and between the saccades they move in a way that is the exact opposite to the movements of the head. Adding the eye and head movements together shows that gaze stays almost stationary between saccades. A simple observation can demonstrate the importance of stationary gaze. Move your finger in front of a patterned background (books, words on a page, or fabric) while tracking your moving finger with your eyes. While continuing to track the finger, look just beyond it at the pattern. All the fine detail will be lost. This is motion blur. It is just like the blur in a photograph when the camera moves, and it is the reason why gaze has to be kept so still. It is the slow response of the receptors that causes the blur: it takes about 20 milliseconds for a retinal cone to respond fully to a brightness change. This may not seem much, but when translated into an image speed it means that detail in the image will start to be lost at speeds above about 1° per second, which is *very* slow. It also means that during saccades, which have speeds of tens to hundreds of degrees per second, information uptake is impossible.

The movements of the eyes that counteract the head movements are brought about by a reflex known as the vestibulo-ocular reflex, in which the head movements are measured by the semi-circular canals of the inner ear, and these measurements are sent to the nuclei in the brain that control eye movements. This reflex is very fast and accurate, and it allows us to maintain gaze on an object despite movements of the head. You can try this out by fixating on something while shaking your head. It is almost impossible to dislodge gaze.

This pattern of saccades and fixations is almost universal in animals with good eyesight. It is seen in all vertebrates, in flies, crabs, and even cuttlefish. In humans and other primates, this eye movement strategy is used to move gaze from one part of a scene to another, to bring the high resolution of the fovea to bear on places from which information is needed. It seems that in evolutionary terms, this is a relatively new use for this strategy, and it was not why it originated. To quote Gordon Walls, a famous pioneer of the study of animal vision: 'Their origin [eye movements] lies in the need to keep an image fixed on the retina, not in the need to scan the surroundings.'[2] Half a billion years ago in the Cambrian, when animals started to see well, the avoidance of motion blur was as important as it is now.

To record human eye movements 'in the real world' (such as those in Fig. 71) requires a particular piece of kit: a head-mounted eye camera that images both the eye and the scene. The history of such cameras goes back to the 1950s, when Norman Mackworth made one based on a helmet-mounted

ciné-camera.[3] This worked, but it was cumbersome, and it was not developed further. By 1990 there were one or two video-based eye cameras, but
they were expensive and had various disadvantages, so I decided to make
my own. This is shown in Figure 72a. It consists of a single, small video camera (c) with a split field of view. The upper two-thirds of the field images the
view ahead via a part-silvered mirror (p), and the lower one-third images a
view of the eye focussed by a concave mirror (m). This produces a video
combining the scene and an inverted view of the eye. Figure 73 shows what
this looks like. The direction of the foveal axis is determined by overlaying
the image of the eye with a computer-generated circle which is rotated over
a sphere concentric with the eye by means of a tracker-ball. This produces an
ellipse which is fitted, frame by frame, to the moving iris of the eye. This
generates the coordinates of a dot which is added to a copy of the original
video, and after suitable calibration, this gives the direction of the foveal line

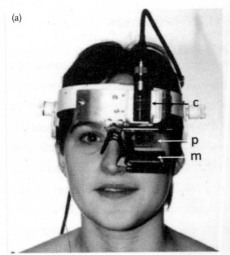

Fig. 72. Two camera arrangements for recording eye movements. a) The camera
described in the text, worn by my colleague Sophie Furneaux. It consists of a video
camera (c) which images the view ahead via a part-silvered mirror (p) and the eye
via a concave mirror (m). b) A more recent spectacle-mounted device which uses
two separate cameras. It is worn by Ben Tatler, who worked on several eye-movement
projects with me, including writing a book on eye movements.[4]

of sight (the white dots in Fig. 73). By the late 2000s, a number of more convenient eye cameras were on the market, including the version in Figure 72b, which uses two tiny cameras fitted to a spectacle frame.

One of the most interesting things about eye movements is that they are not available for conscious scrutiny. Rather like breathing, they *can* be directed by conscious control—you can decide to look at something—but for 99% of the time they get on with the job of directing the eyes to places from which information is needed. They leave no trail; you know where the current fixation is directed, but you can only guess at the location of the previous one. This is what makes eye movement recording particularly exciting; it is the only way of finding out what our eye movement system gets up to. In the sections that follow, I will look at three complex tasks—driving, reading text and music, and playing ball sports—in each of which eye movements play a vital, and often unexpected, role.

Where Do We Look when Driving Round a Bend?

The first use of our eye camera was on a project to work out where the eyes are directed when steering a car on a winding road. My thinking was that turning the steering wheel by the right amount requires specific information about the upcoming road curvature, and that finding out where the eyes look might tell us where this information comes from. I collaborated with Dave Lee, of the Psychology Department in Edinburgh University, who had managed to acquire a Jaguar car that had been used in another motoring project. The car was instrumented, and had a computer in the back which registered a variety of aspects of the car's performance including, crucially for us, the steering wheel angle. We thought that by comparing gaze direction with steering wheel angle we might find whether they were related, and if so how.

We drove the car round Queens Drive, a road that winds round the base of Arthur's Seat—the famous rocky outcrop that dominates Edinburgh. It was a one-way road, which avoided the problem of oncoming traffic, and it had many bends in both directions. Four of us drove the 1-km course

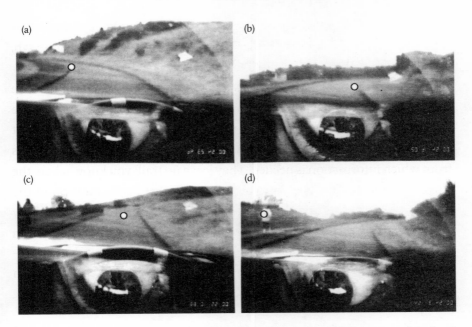

Fig. 73. Representative frames from the eye-camera video of a drive round Arthur's Seat in Edinburgh. The top two-thirds of the frame shows the road, and the bottom one-third is an inverted image of the driver's eye. The position of the iris relative to the head is used to generate the location of the dot which shows the direction of the fovea on the upper part of each frame. In the left-hand bend in a) and the right-hand bend in b), the eye direction meets the road at a point where the line of sight is tangential to the inside of the bend (the tangent point). c) On a straight stretch, the eye views the centre of the road, and when there is something more interesting to look at (a jogger), the eye can disengage from the road for a second or two.

wearing the eye camera. When we inspected the eye camera videos afterwards we saw immediately that, when entering a bend, we always looked at a specific part of the road edge, known as the tangent point (Fig. 73a and b). This is the point where gaze direction is tangential to the edge of the inside of the bend. It is not a fixed point but moves round the bend with the car.

We had expected some sort of relationship between gaze direction and steering, but when we looked at the steering wheel data from the car's computer and compared it with the gaze direction data from the eye camera, the result amazed us. Plotted against time the two records were almost identical (Fig. 74a), and this suggested that the direction of gaze might in some way be

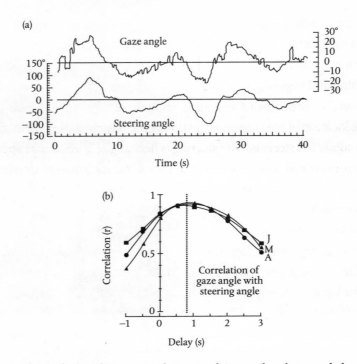

Fig. 74. a) Plots of gaze direction, relative to the car's heading, and the steering wheel angle for a 40-second segment of the drive around Arthur's seat. The two graphs are almost identical. b) Correlation between the two graphs in a) for different delays. The best fit occurs when gaze leads steering wheel angle by 0.8 seconds.

the cause of the amount driver turned the steering wheel. The conclusion seemed to be that you steer where you look.[5]

If gaze direction changes do indeed *cause* steering changes, there should be a delay between the gaze angle and steering wheel angle (certainly not the other way round!) We could test this by cross-correlating the two graphs. That means moving one graph along the time axis relative to the other in small-delay steps and finding the delay at which the correspondence between the two graphs is greatest. The result is shown in Figure 74b. The correlation between gaze direction and subsequent steering wheel angle is maximum when the delay is about 0.8 seconds, from which we can infer that this is the time it takes to convert gaze direction into action. Another way of looking at it is that 0.8 seconds should be the time it takes for the car to reach the tangent point which, at a legal 30 mph, will be 35 feet or 10.6 metres ahead. This is about right for the bends on this road.

As we have seen, drivers look at the tangent points on the road edges when approaching and negotiating bends. (On a two-lane road, one of the tangent points is on the centre line.) This suggests that there may be something special about the location of the tangent point that gives the driver the information he needs to adjust his steering. The steering wheel angle, at moderate speeds, is directly proportional to the curvature of the car's track (equal to $1/r$ where r is the radius of the bend). So we can put the question another way: is there a simple relationship between the driver's gaze angle (θ in Fig. 75) and the curvature of the car's path? The answer is yes. As Figure 75 shows, the curvature of the bend, $1/r$, is given by $1/r = \theta^2/2d$. (The geometry is given in the legend to Fig. 75.) This means that the steering wheel angle, which determines the curvature of the car's path, is predicted by two numbers: the angle of the driver's direction of gaze θ, and d, the distance of the driver from the edge of the lane.

We can assume that θ is known implicitly to the driver because it is his gaze angle—the sum of his eye-in-head and head-on-trunk angles. His trunk will not move relative to the car if he wears a seat belt, so θ is also the angle between the car's heading and the tangent point. To convert θ to path curvature also requires d, the distance from driver to the lane edge. In normal driving this is likely to be fairly constant, but in a separate experiment with a driving simulator, we found that while the fovea is looking at the tangent point, the lane edges close to the car are being monitored by peripheral vision, so that both θ and d are available to the driver. Our conclusion was that when driving on winding roads, drivers adjust their steering by directly measuring their gaze angle relative to their body axis, while looking at the tangent point on the inside of each bend. Many studies have confirmed that drivers monitor the tangent point on bends, but not all agree with us that the direction of the tangent point is the main variable that makes steering possible. It is true that it is not obligatory to track the tangent point all the time, but when drivers are instructed to hold their gaze away from the tangent point, their steering becomes more erratic.

Of course, driving involves much more than just lane keeping. There are road signs, other vehicles, pedestrians, and unforeseen hazards, all of which

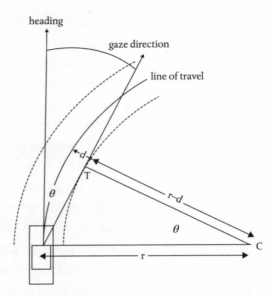

Fig. 75. The geometry of steering round a corner. The driver views the tangent point T at an angle θ from his current heading. θ is also the angle between the driver, the centre of curvature of the bend C, and the tangent point T. The driver's line of travel is an arc centred on C, whose radius is r, and the inner radius of the bend is r–d, where d is the distance of the driver from the edge of the road. From ordinary trigonometry $\cos \theta = (r-d) / r$, or $r = d / (1-\cos \theta)$. $\cos \theta$ is known to be equal to $1 - \theta^2/2$. Substituting this expression for $\cos \theta$ then gives the curvature of the bend $(1/r)$. $1/r = \theta^2/2d$.

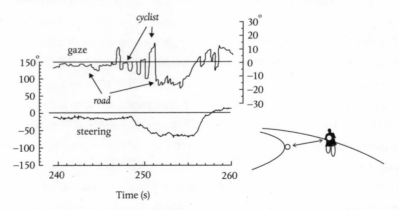

Fig. 76. Driver and cyclist. The driver's gaze direction switches between the tangent point (as in Fig. 75a) and the cyclist every half-second. The steering follows gaze only when the driver views the tangent point, and is disconnected from gaze when monitoring the cyclist.

require diversion of attention away from the road edges for varying periods. A nice example of this is shown in Figure 76, in which the driver, on a left-hand bend, intermittently monitors a cyclist on the right. Comparing the gaze and steering records, it is clear that steering is only related to gaze when the driver views the road edge, and this connection is broken every time the driver views the cyclist. Just as well for the cyclist!

Reading Words and Music

As early as 1920, Guy Buswell, working in Chicago, recorded the eye movements that accompany the human voice while reading aloud. This was quite a feat. He used a microphone of his own design and an eye camera in which a spot of light, reflected from the subject's cornea, wrote directly onto moving photographic film. This was decades before 'head-free' eye cameras were in use (Fig. 72), and Buswell's subjects had their heads firmly clamped in the apparatus. The results of one of these early recordings is shown in Figure 77.

Buswell found that the eye–voice span in reading aloud—the time between looking at a word and uttering it—was on average 0.79 seconds, or about thirteen letters. That is the time it takes to recognize a word, form it into an appropriate set of muscle contractions, and speak it. This is a rolling process, rather like a production line: letters enter the eye, pass through the brain machinery, and are then ejected as the word is spoken. At some stage the meaning of the passage is extracted, but not on a word-by-word time scale. Reading is rarely a simple succession of left to right of fixations (or right to left or top to bottom, depending on the language). Regressions are common, as in the first three fixations in Figure 77, where fixations 2 and 3 precede 1 on the line.

Reading has been studied extensively since Buswell, and it has become clear that the details of where and when saccades are made, and the durations of fixations, are affected by many factors. Some are not to do with the meaning of the words themselves. Word length, word spacing, and where on a word the previous saccade landed all affect the size and timing of the next

Fig. 77. Buswell's historic record from 1920 of reading aloud by a high school freshman (about 14 years old). The vertical lines are fixations. The upper numbers give the sequence along each line, and the lower numbers the fixation durations in units of 0.02 seconds. The dashed lines show the eye–voice span V–E, E being the gaze direction while the word at V was being spoken.

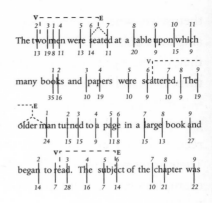

saccade. Others factors are related to the difficulty of the words themselves, with unfamiliar words resulting in longer fixations. An important technique for teasing out these factors has been gaze-contingent masking: a method in which changes are made to the text *during* saccadic eye movements, so making them invisible to the reader. If the seven letters centred on the fovea are always masked out the reading rate reduces massively (from 250 words per minute down to 12), but masking everything except those seven letters has little effect on reading rate. This again emphasizes the crucial role of the fovea. Characters up to fifteen letters ahead of the fovea on the line can affect the sizes and landing points of subsequent saccades, but they are not read as letters.[6]

Reading piano music is a skill that takes years to master. A good pianist can take in two lines of music from the treble half of the stave and another two lines from the bass part, and then 'emit' them simultaneously in the form of movements of four fingers. Like reading text, this is a rolling process, with note sequences continuously entering the eyes, being transformed by the brain into motor commands, and then turned into muscle movements. You might think that some instruction in developing an appropriate eye movement strategy would be an important part of music training, but this is not so. There may be exhortations to keep ahead of the notes being played, but not how to move gaze to acquire the notes themselves in an efficient way. If you ask pianists where they look when sight-reading, they usually do not know. Pianists who read this might like to ask themselves whether they move their gaze along the gap between the upper and lower parts of the stave, taking in the notes simultaneously, or whether they alternate gaze, taking in clusters of notes from treble and base clefs in turn. They will probably find this hard to answer.

The first person to tackle this question was Homer Weaver, in 1943.[7] He got experienced pianists to play musical pieces they did not know, and recorded their eye movements with an apparatus similar to Buswell's. Their heads were restrained, so it must have been quite uncomfortable. An example of such a recording, of a minuet by Bach, is shown in Figure 78a. Both Weaver's records, and our own from half a century later, show that

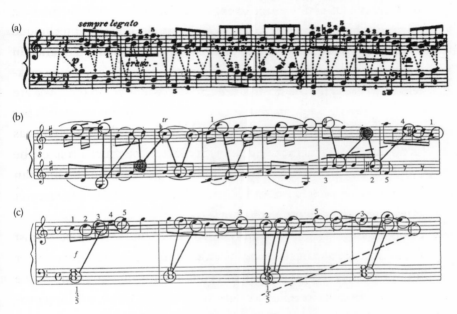

Fig. 78. Sight reading music. a) Weaver's recording, made in 1943, of a competent pianist playing a Bach minuet. The dotted line shows the track of the eye. Gaze alternates between upper and lower halves of the stave, taking in one to three notes with each fixation. b) An excerpt from our own study in 1999 of a professional accompanist sight-reading a Scarlatti sonata (presto). Circles represent fixations and are 1° across. As in a), the pianist alternates between the two parts of the stave, taking in several notes with each fixation. The filled circles represent glances down to the keyboard. The pianist does not return to the same point, but to the next fixation point in the sequence. c) A novice playing 'The five fingers' exercise from Czerny. In contrast to a) and b), almost every note is fixated, sometimes more than once.

pianists do not try to view the whole stave from a central corridor between the upper and lower halves, but alternate fixations between the two. This makes the processing task even more complicated than in text reading because the treble and bass notes of a chord are acquired at different times but have to be paired up and emitted simultaneously as finger movements.

My colleague Sophie Furneaux studied pianists with different levels of expertise.[8] One obvious difference was that novices were much slower, and they tended to look at every note, often more than once (Fig. 78c). Interestingly the 'eye–note span'—the time between viewing a note and playing it—changed little as pianists got better, and it remained at an average of roughly 1 second. Other things did change. The number of notes within that span—the number being processed through the brain at any one time—was about two for novices and four to five for professionals. Secondly, the variation in the eye–note span was almost constant for professionals, between 0.5 and 1.1 seconds, whereas for the novice, it fluctuated wildly between 0 and 2.6 seconds. As a novice who failed to improve, this confirmed my belief that piano playing consists mainly of mistakes with short bits of tune in between.

As skill develops, whether in reading, typing, or sight-reading music, speed and fluency are increased by dealing with groups of words or notes together, rather than singly. This is known as 'chunking', and rather like predictive text, it involves recognizing phrases or runs of notes that often occur together. Something like this seems to be happening in the left-hand part of Figure 78b, where runs of three notes only require a single fixation. As a piece becomes more familiar, chunking becomes more obvious. Parts of the score become available from memory and may only require a visual prompt to bring them into the processing stream.

Ball Sports

In cricket a fast bowler can propel the ball at 90 mph, meaning that it reaches the batsman in 0.4 seconds. In that time the batsman has to assess when and where the ball will reach his bat, and prepare a stroke that will propel the ball in a useful direction. The situation is depicted in Figure 79, which shows a famous batsman, Victor Trumper, about to make a scoring stroke (a straight drive) at the Oval cricket ground in 1902. The box in front of him represents the place and time that contact must be made with the ball if the shot is to be successful. It is 5 cm high, and 5 *milliseconds* long. The batsman's timing needs to be exceptional.

The short time the ball is travelling from bowler to batsman is long enough for the batsman to make one saccadic eye movement. Where should he direct gaze to get the information needed to predict the ball's time and place of contact with the bat? Peter McLeod, a psychologist at Oxford University and a keen cricketer, was interested in this question, and he invited me to help. We suspected that where the batsman looked would tell us something about where he was getting the information needed to coordinate the contact between bat and ball (Fig. 79). In cricket the ball (usually) bounces once before reaching the batsman. From the batsman's point of view, seeing what the ball does when it bounces is important, not only because the ball may change direction, but also because it provides key information about where and when it will arrive at the bat. This is illustrated in Figure 80.

Peter had lined up three batsmen of different abilities: a professional county player, the university coach, and a distinguished academic who played on Sunday afternoons. The cricket club had indoor nets and a bowling machine that could be set to generate deliveries of different trajectories and speeds. For reasons of prudence, and the safety of the eye camera, the

Fig. 79. The batsman Victor Trumper about to make a straight drive at the Oval cricket ground in 1902. The box shows where the ball needs to be at the time of contact with the bat. It is 5 cm high and represents 5 milliseconds in time.[9]

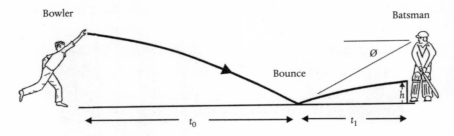

Fig. 80. Information available to a batsman. To make a useful stroke the batsman needs to know the time from the bounce to contact (t_1) and the height of the ball when it arrives (h). What he can measure are the time the ball left the bowler's hand to the time it bounced (t_0), and the angle down from his eye to the bounce point (ϕ). The bowler can vary the speed of the ball and the position of the bounce point, so t_1 and h will both vary too. Seeing where and when the ball bounces is of crucial importance to the batsman. The task of the batsman is to convert t_0 and ϕ into t_1 and h.

speeds did not exceed 'medium pace'. Even with the headgear, all the batsmen were able to hit the ball reliably.

Given the importance of the bounce point, we thought that the batsmen would move their eyes to get a good view of the part of the pitch where the ball bounced. This turned out to be true, nearly every time. You might think that batsmen would simply follow the ball as it descended towards the pitch. This is, after all, what coaches tell you to do: *keep your eye on the ball*. This is not what our batsmen did (Fig. 81). They looked closely at the delivery of the ball—the point at which it emerged from the bowling machine (1 in the figure)—and waited for a minimum of 0.15 seconds, as the ball descended a few degrees in their field of view (2). What happened next did surprise us (and them). They made one saccade to a point on the pitch close to where ball was expected to bounce (3). For the two better players this meant that they had time to observe the bounce, giving them the two measurements they needed (t_0 and ϕ in Fig. 80) to estimate when and where the ball would arrive at the bat. The record in Figure 81 shows the descent of the ball, taken from the view in the eye camera, and the batsman's gaze direction over the same period. In this recording gaze arrives very close to the bounce point, about 0.16 seconds ahead of the bounce itself; the upper record shows that this involves taking gaze to a

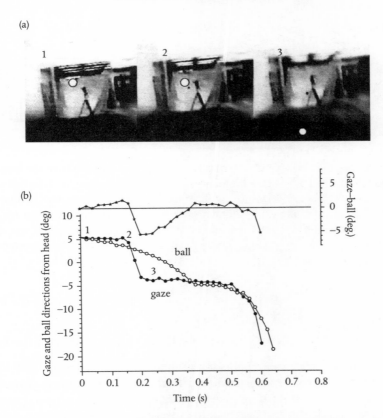

Fig. 81. Gaze movements of a batsman watching a delivery from a bowling machine. a) Frames from a video of the view from the eye camera on the batsman's head. (1) Just before the delivery; gaze (o) is on the bowling machine. (2) About 0.1 seconds later; the ball (•) has dropped a few degrees in the field of view with gaze beginning to follow. (3) The ball continues to descend, but gaze has now moved to the estimated bounce point with a single saccade. b) Lower record shows the vertical direction of the ball and the gaze point as seen from the batsman's head. The saccade arrives at the bounce point about 160 milliseconds before the ball, and, thereafter, gaze tracks the ball up to contact with the bat. The upper record shows the angular difference between gaze and ball.

point 5° away from the ball. After the bounce (the inflection in the ball's trajectory) gaze follows the ball onto the bat, until contact is made about 0.65 seconds after the delivery from the bowling machine.

There were some interesting differences between the three batsmen (Fig. 82). Mark, the best batsman, only made a complete saccade to very

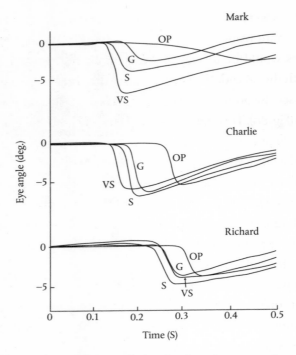

Fig. 82. Vertical eye movements made by the three batsmen when watching the approaching balls of different delivery length. VS, very short, bouncing nearest to the bowler; S short; G good, the ideal length; OP, over-pitched, bouncing nearest to the batsman. Mark, the best batsman, used saccades for short balls but smooth tracking for over-pitched balls; Charlie always used saccades, as did Richard, but Richard's saccades had latencies that were 0.1 seconds longer than the better players. The eyes all move up after the saccade reflecting the downwards movement of the head, unlike the gaze records in Figure 81, where head and eye movements are combined.

short deliveries: these are balls that bounce closest to the bowler and so need a very quick response. For the others he employed a mixture of both a saccade and smooth tracking. With the over-pitched balls, which bounce closest to the batsman and so moves downwards only slowly in the field of view, he used smooth tracking to follow the ball all the way. Smooth tracking is accurate but slow to start and so can only be used for objects that move slowly. Charlie, the competent amateur, used saccades every time, but with lengthening latencies as the bounce points moved up the pitch towards the batsman. Richard, the weakest player, was similar, except that the saccade latencies were almost 0.1 seconds longer that for the other two. This meant that Richard would have difficulty facing a fast bowler, as his downward saccades would be too late to register the bounce or if the bounce coincided with the saccade, he would not see them at all. Assuming that the batsman can see the bounce point, he now has the task of converting t_0 and ϕ into t_1 and h (Fig. 80). This can be done, as Peter and I showed mathematically in our paper.[10] In practice a batsman has to learn this transformation by trial and error, and it takes years of practice to become good at it. There are some tweaks that need to be made on the day of the match: the heights of the bowler and batsman and the springiness of the pitch all have small effects on the conversion. Most batsmen take a while to 'play themselves in' before trying to make attacking shots; presumably, this gives them a chance to assess the bowler and the pitch.

The most interesting outcome of this study was that batsmen made saccades to points on the pitch where there was nothing to see (except grass). It is usually assumed that eye movements are made towards objects of interest in the peripheral field of view, but in this case, there were no objects there to catch the eye. Evidently these saccades are being made to points from which information will be needed in the near future: they are anticipatory rather than reflexive and are made on the basis of previous experience.

We found it amusing that within days of the publication of our paper, we had requests for information from both the Australian and the West Indies cricket boards, but nothing from our own in Britain. I suspect they still believe batsmen should keep their eyes on the ball.

In another sport, squash, the role of eye movements in anticipating the future path of the ball is particularly impressive. Squash, like tennis, is played between two contestants, and the aim is to bounce the ball off the front wall on a trajectory that makes it hard for the second player to return it. The ball can bounce off one or many side walls, but not more than once on the floor, before being returned to the front wall. Figure 83 shows a single example of a return shot showing the path of the ball and the eye movements of the player about to return the ball. It was filmed in 2008, by Travis McKinney and his colleagues in Austin, Texas. Even after my experience

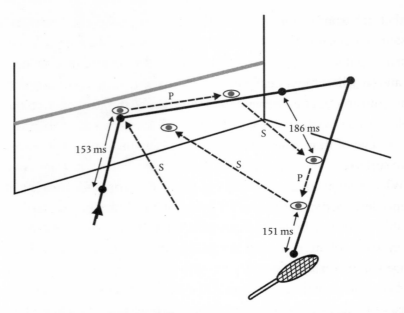

Fig. 83. An example of the return of a ball by a squash player. The full lines and dots show the path of the ball, which bounces off the front wall and then the right side-wall. The eye symbols and dashed lines show the path of the eye. P, pursuit; S, saccade. Further explanation in text.[11]

with cricketers, I was still astonished by the extent to which, in squash, the eyes managed to anticipate the future path of the ball. Although impressive, this pattern of eye movements is far from exceptional. In what follows I will try to give a commentary on the events shown in the figure.

The ball arrives from an unseen player on the left. The returning player (with the racket) sees the ball coming and makes a saccade close to the bounce point, arriving 153 ms before the ball (1). For a short time, the eyes pursue the ball smoothly; then, while the ball is still moving towards the side wall (2), they make a saccade to a point where the ball will arrive 186 ms later (3), *after* the ball has bounced off the wall. Following another short episode of smooth pursuit, the eyes saccade (4) to a point on the front wall where the returner expects the ball to arrive after it has been hit (5). This saccade occurs 151 ms *before* contact is made with the ball. What this shows is that, far from simply following the ball around, the player's saccadic system anticipates the ball's future position by 100–200 ms, much as the batsmen did in cricket. Particularly interesting here is the saccade between (2) and (3), where the system is not simply extrapolating an existing path but takes into account the change of direction caused by the bounce off the side wall. This shows that the saccadic system has at its disposal a working knowledge of those laws of physics that apply to the interaction of balls with surfaces.

What general conclusions can we draw from the patterns of eye movements described in this chapter? What surprised us most was that we were unable to predict, by introspection or intuition, where our subjects were likely to be looking. It was not obvious, a priori, that drivers would fixate the tangent point on bends or that pianists would alternate gaze between upper and lower parts of the stave or that batsmen would look at the bounce point ahead of the ball's arrival. These were all discoveries made possible by recording eye movements. What they tell us is that the eye movement system has its own knowledge base, which it does not share with the regions of the mind that allow conscious scrutiny.

6

THE MIND'S EYE

The World Out There and the World in Your Head

Human vision is an extraordinary field, spanning philosophy, psychology, and physiology. Central to thinking about our vision has been the disturbing realization that the world is both 'out there' and in one's head at the same time. In this chapter we will explore some of the many intriguing ideas and paradoxes of human vision.

When I look out of my window, in one direction, I can see my garden. The flowers are of many colours against a background of green foliage, with trees in the distance. In the other direction, there is a road with fast-moving traffic and a background of houses. That there is a real world out there is beyond reasonable doubt, but for me to see that world, it has to be processed through the machinery of my brain. My view is a subjective view, unique to me. The relationship between my view and the objective outside world, which seems so straightforward, is in reality extremely complex. This problem—the relationship of the subjective and objective worlds—is often referred to by philosophers as the 'hard problem' of consciousness. There are many aspects to this problem, but I will deal with the three that have particularly interested me. What is the nature of colour? How does diverse information about the distances of objects, extracted by our visual system, become a reconstituted three-dimensional outside world? And how are our jumpy retinal images converted into a seamless view of the world?

Consciousness and Colour

Let me begin with a classic example: the redness of red. In the objective world, there is no colour, just different mixtures of wavelengths of light which arise because different surfaces reflect wavelengths selectively. The wavelengths important to humans lie between 400 and 800 nanometres (1 nm = 10^{-9} metres) on the electromagnetic spectrum. We see the longer wavelengths as reds and the short wavelengths as blues. Bees can see shorter wavelengths (down to 320 nm in the ultraviolet), and some snakes make use of longer infrared wavelengths to track warm-blooded prey. Photons of light are not coloured in any useful sense. The three types of cones capture photons and separate them into three wavelength ranges. Colour itself only arises much later in the visual pathway. The captured photons are converted into electrical signals, which, after various transformations, reach a very specific region of the visual cortex known as visual area 4 (or V4). This is one of a number of regions that surround V1, the primary visual cortex and the site of entry for most of the information reaching the cortex from the eye (Fig. 84). These various regions deal with different aspects of image information, such as depth, motion, shape, and colour.

To say that subjective colour is *produced* in area V4 is a very strong statement. Semir Zeki, in the 1980s, made microelectrode recordings from the brains of monkeys, and identified part of V4 as concerned with colour (parts of V4 also deal with shape).

The most striking and revealing proof that colour is localized to one region of the brain of humans comes from an account in 1987 of a colour-blind painter, Jonathan I, told by Oliver Sacks and Robert Wasserman.[1] Jonathan I was a successful 65-year-old artist who specialized in colourful abstract paintings. He was involved in a car accident, and was taken to hospital with concussion. When he came to, he found that he was unable to recognize letters or see colours. The letters came back in a few days, but he remained completely colour blind. Everything became monochrome, and although his black and white vision was as acute as ever, there was no hint of colour. He found this really distressing. Flesh tones, of himself and his

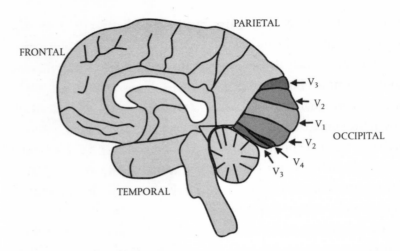

Fig. 84. View of the *medial* face of the right cerebral hemisphere, on which the different visual regions are visible. This is the view from the central cleft between the hemispheres, looking outwards. Very little of this is visible from the outer (lateral) surface of the cortex. Nearly all information from the eyes arrives at V1, the primary visual cortex, in the occipital region of the cortex. This is surrounded by the secondary areas V2–V4, most of which lie within the cortical folds. Other areas such as V5 are not visible in this view. The areas differ in function. *Very* roughly, V1 classifies the incoming information according to the orientations of edges; V2 continues this, but it is also involved in analysing depth through binocular differences (disparity); V3 neurons respond to motion of the whole field of view; V4 is responsible for colour and shape; and V5 (not visible) deals with the motion of objects. The four main cortical regions (or lobes) are also labelled.

wife, became 'rat-coloured', and people were like 'animated grey sculptures'. Food was grey and disgusting. Tomatoes were black. Flowers had no colour, and visiting art galleries was no longer a pleasant experience. What was particularly striking, and intriguing from our point of view, was that he could not *imagine* colour. It was as though the piece of brain machinery that produces colour, real or imagined, had been destroyed. Mr I became depressed, and over a period became nocturnal, walking the streets at night when monochrome vision made more sense. He took up painting again, but without colour.

Jonathan I's condition is known as cerebral achromatopsia. It is very rare, because it is unusual for a haemorrhage to affect both sides of the cortex symmetrically and to such a limited degree, that only colour is affected. The damage in Mr I's case was so limited that it could not be located by the scanning systems of the 1980s. One-sided cerebral achromatopsia is rather more common—half the visual field loses colour. Either type of central colour blindness is quite unlike ordinary colour blindness, which arises from a genetic malfunction of one or more of the three visual pigment types in the cones of the eye. In fact, tests on Jonathan I showed that he had a perfectly normal complement of visual pigments. What he had lost was the ability to manufacture colour from the signals from the eye that encode the three wavelength channels.

Intriguingly, none of the other attributes of Jonathan I's vision was affected. Brightness, shape, speed, depth, and object identity were all intact. If consciousness were a unitary state involving the whole brain, this couldn't happen. Zeki explains the apparent unity of consciousness by assuming that each cortical region (or *node*) has its own 'microconsciousness' and that these combine, via a network of lateral connections, to provide the complete percept that we have, for example, when looking out of a window.[2] Many neurophysiologists working on vision are inclined to accept some version of Zeki's view. What Zeki's argument fails to deal with is the nature of 'qualia': the subjective qualities of percepts. Besides red and other colours, qualia would include the sound of trumpets, the smell of honeysuckle, or indeed, the experience of pain. To be fair, no other proposal seems to have made much progress with this ancient philosophical problem. Qualia are private to the perceiver, and very resistant to objective study. It may be that we simply have to accept that qualia are the ways the workings of parts of the brain present themselves. I am tempted to say 'to us' but that isn't right. In this context our brains *are* 'us'.

The alternative to this idea that consciousness arises at some 'intermediate' level in the brain, in areas such as V4, is that somewhere higher up in the brain, there is a region where all the components of the scene analysed by the various cortical regions around V1 come together into a single unified percept, and this is what we 'see'. Intuitively this seems appealing; there is a 'me' looking out at the world, which seems to be a single vista, not split up into component attributes. This view of consciousness is usually attributed to René Descartes (1596–1650), who held what seems to us now to be a pretty weird idea that consciousness requires an immaterial soul (aka. mind) which interacts with the body via the pineal gland in the brain. (His reason for choosing the pineal gland is that it is one of the few parts of the brain that is central and not bilaterally duplicated.) For Descartes, this bottleneck between sensory input and motor action is where consciousness occurs. Daniel Dennett, in *Consciousness Explained* (1991), recasts Descartes' idea as a theatre in which a 'homunculus' observes all the sensory data on a screen, and from this makes decisions and sends out commands.[3] Dennett, unsurprisingly, rejects this idea for one in which consciousness arises simply as a consequence of the complex flow of information in an active brain. Dennett's book has caused much controversy among philosophers, not least because it does not really distinguish between a conscious human, and a very smart automaton that behaves as a human (a 'zombie'). However, perhaps the most cogent argument against the 'screen' idea is that no such region has been found. Writing in 2011, the vision scientist Bruce Bridgeman put it this way:

> Physiologically, we were already certain by then [1994] that no brain area contained a panoramic, high acuity representation that corresponds to our perceptual experience.[4]

While Bridgeman's statement is certainly correct, there has to be, somewhere, a representation of the visual environment that is both broader than the current field of view, and which has our eye movements edited out. Beginning in about 1998, the use of scanning methods has shown that there are three areas of the cortex (the 'scene network') that have at least some of the attributes that would be needed to construct such a coherent world picture. These areas are the *occipital place area* (OPA) on the outer face of the cortex above the main visual areas, and two areas on the medial face: the *parahippocampal place area* (PPA), and the *retrosplenial complex* (RSC) (Fig. 85). In a study by Caroline Robinson and her colleagues,[5] subjects were shown a full 360° panorama of a scene and were then tested to see whether or

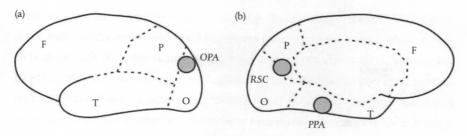

Fig. 85. Three areas in the cortex (the 'scene network') involved in the recognition of scenes. a) is an outside view of the left hemisphere showing the location of the occipital place area. b) is a medial view, showing the retrosplenial complex and the parahippocampal place area. The four lobes (occipital, parietal, frontal, and temporal) are indicated with their approximate boundaries.[5]

not they could identify small segments of the scene as coming from the same panorama. That they could do this indicated that visual experience forges memory associations between different parts of a panoramic scene. When tested in an fMRI scanner, two regions were uniquely involved in establishing this memory, the OPA and RSC. The PPA was not involved in panoramic memory, but was concerned with the visual recognition of landmarks of relevance to navigation. The authors of the paper were careful not to claim that these areas 'contained' a panoramic scene, but they concluded that 'ongoing scene representations are affixed to a broader representation of the surrounding environment, which may help to support our sense of a seamless panoramic visual experience'. The present consensus seems to be that the current retinal image, analysed by all the different regions in the occipital lobe and elsewhere (Fig. 84) is what we see, but other parts of the cortex cooperate to situate this view in its appropriate remembered setting.

Putting the World Back 'Out There'

An eye, like a camera, produces an image which is a two-dimensional projection of a three-dimensional scene. The left–right and up–down directions of points in the image can be read off directly from the image. Although somewhat distorted, this 'retinotopic' geometry is maintained in all the early stages of the visual pathway in the cortex. Distance information is different. It is not directly available from the image the same way as the other two dimensions, and it has to be derived from a variety of clues contained in the images from the eyes. Yet, as far as our conscious percept of a scene is concerned, the depth dimension is just as persuasive as the other two.

Depth clues are plentiful. For distances up to a few metres, the most reliable information comes from disparity: the differences between the images in the two eyes. It is easy to see these differences by focussing on your finger, alternately shutting one eye or the other, and seeing what happens to objects in the background. A simple example of how this works is shown in Figure 86. Here two outline views of a pyramid with its top cut off have been drawn as they would appear to the two eyes. Squinting at this picture so that the two images fuse to give a single central image, flanked by the two separate images, gives a very strong impression of a three-dimensional structure. A more famous example of depth from pure disparity is the random-dot stereogram devised by Bela Julesz (Fig. 87). Here two squares of random dots are presented separately to the two eyes. Most of the dots are identical in the two squares, but in the centre of each is a smaller square,

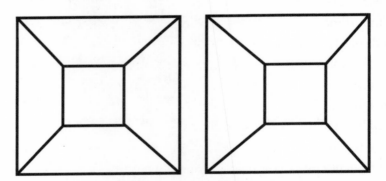

Fig. 86. A pair of drawings of a truncated pyramid as seen from the separate viewpointsw of the two eyes. To view this in depth, squint so that the images cross to form a single combined central image flanked by the single images. The fused central image will appear as a three-dimensional object. Note that images from the left and right eyes have been reversed to allow fusion by squinting. It may take a little practice to fuse the images, but it is worth it!

Fig. 87. Pair of random-dot stereograms, with a displaced central square. To see this in depth you have to squint, as in Figure 86, so that the two squares fuse to give three images, the central one being the fused image. In that central square a smaller square will appear, hovering in front of the main square.

where the dots have been displaced to the left or right. These are undetectable to casual observation, but when seen together, by fusing the two images, the smaller square stands out as being in a different plane, about half a centimetre in front of the background. A similar principle applies to 'magic eye' pictures in which two images are present in the same image, and by relaxing accommodation, each can be seen separately by the two eyes. The importance of these demonstrations is that depth can be determined from disparity, independent of the objects in a scene. Exactly where in the cortex the images from the two eyes are combined to extract depth is still not clear. Cells with some of the right properties have been found in area V1, and neighbouring areas V2, V3a, and V4.

In addition to this 'objective' clue, many others derive from 'pictorial' features of the image. The Italian artist Canaletto (1697–1786) was a master at inducing depth by exploiting these clues. In his picture of the Grand Canal in Venice (Plate 8), we can see many of them. First and foremost is perspective: the way that parallel lines converge on a vanishing point in the distance, with the amount of convergence at any level indicating its distance. Related to this, the texture of the ripples in the water appear coarser in the foreground. From the high viewpoint here, nearer figures are lower in the scene.

The men in the boats appear smaller when they are distant than when they are near, and as the height of a man is fairly predictable, his size in the image is another depth cue. Nearer objects occlude more distant ones, which makes it possible to rank the figures in the boats in their order of distance. Really distant objects may appear in lower contrast than nearer ones because of atmospheric haze (although Canaletto preferred to paint scenes with clarity and high contrast). The combination of these cues *almost* defeats the powerful information coming from the disparity mechanism that this picture is actually flat.

Physiologically, what one would like to find is an area in the cortex which responds to all these clues, as well as disparity, to provide a 'distance node', which, just like Zeki's colour node, can make its contribution to the subjective reality of the scene in front of us. There have been several ideas about where this might be. It has long been known that area V5 (also known as the middle temporal area [MT]) is concerned with the motion of objects, but a recent suggestion[6] is that this motion generates depth information. When you move through the world, the speed that objects move across the retina depends on their distance: far objects move slowly, and near objects fast. Because of this, a map of object motion in the world appears, to a moving observer, as a map of inverse distance. In this way MT neurones can act as distance monitors. Some MT neurones also signal depth from disparity, the other accurate source of distance information. So, although it cannot yet be said that MT is the 'distance node', this is beginning to look probable.

The other great problem with depth is that, although it is concocted in the brain from a mixture of clues, it is not seen as being inside the head (how could it be?) but back in the world, where the structures in the scene really are. It is as though the three-dimensional contents of the conscious image were somehow being projected back out of the brain into the world. (Some ancient Greeks thought this was literally happening, but physics does not allow such mystic stuff.) A familiar situation in which sensation appears to operate at a distance is when you prick your finger. The painful sensation, we know, originates in the brain, but that is not where it is felt. One might argue that the sensory nerves in the finger are in some sense an outpost of the brain, but this does not explain the kind of referred pain that amputees sometimes feel. The pain from the phantom limb remains 'out there', where it would have been in an intact limb, even though there are no longer any nerves connecting the site of the pain to the brain. This example of the attribution of a sensation to a source to which it has no direct connection is not, in principle, different from seeing distant objects as distant. Rather like colour, making three-dimensionality from clues derived from the world is a trick of the mind whose basis we seem unable to pursue in any satisfying way.

Maintaining a Stable Image of the World

When we look out of the window, we might be tempted to think that what we see is just a combined and lightly massaged version of the images from the two eyes, but it is much more than this. In many ways it is much more user-friendly.

We saw in Chapter 5 that good resolution in each retinal image is limited to the fovea and a region a few degrees around it. By 20° from the centre of gaze, resolution is down to one-tenth of its central value. If you fixate a word on this page and then try to guess the identity of the word two removed to the left or right of the target word, you will probably not be able to do it. The same would apply to the titles of books on a shelf or notes on a musical score. It takes some effort to keep the eyes still enough to observe objects away from the direction of gaze and appreciate just how bad peripheral vision really is. Of course, we overcome this limitation by moving our eyes, and with them the foveas, several times a second. Subjectively we see a world that always appears sharp because it is the fovea's view that always wins out over the indistinct periphery.

But each time we move our gaze to a new location, we change the image on each retina. We do not see either the motion induced by these shifts or the resulting spatial shift of the scene. Unless we pick out specific objects to look at, it appears to us that we are simply moving our gaze smoothly across the view. Various mechanisms are known that may stop us seeing the movement itself. Saccadic eye movements are fast—hundreds of degrees a second—which would blur out any detail during the movement. There is also a specific mechanism, 'saccadic suppression', which partially blanks out vision during each saccade and ensures that the movement itself is not intrusive. The shift in location, however, is harder to deal with. If you make a film that reproduces the shifts in a scene that occur during an eye-movement sequence, those movements are not only visible to a viewer, and they are quite disturbing. We can conclude that image shifts during normal viewing are dealt with by the brain in some way that renders them invisible.

A venerable idea, first proposed in the nineteenth century, is that when an eye movement is made, a copy of the instructions to the eye muscles is sent back to the parts of the brain that deal with the image to let them know what is happening. With this 'efference copy' signal (also known as 'corollary discharge'), these regions can then, in some way, compensate for the actual image motion. In simple terms, they could cancel out the real motion by subtracting the efference copy from it, so creating an image in which nothing has moved: a stable world. It is an attractive idea, and there is now strong evidence that there is a pathway in the brain that provides efference copy from the superior colliculi in the mid-brain (which execute the eye movements) back to the frontal cortex.[7] What is unclear, at least to me, is what it is that needs compensating. When I look up, for example, to look at a bird in a tree, I look *up*: there is no sense in which the location of this upwards-gaze shift has been cancelled or compensated for. On the other hand, there seems to be an underlying world in which I walk, interact with objects, and, in particular, make decisions about where to look next. This world of planning and action is, in some way, fixed in the outside world, and it does not move with the eye. Presumably it is this world that *does* require compensation when it derives information from the visual world. I want to retain the idea of this dual scheme, in which the visual world of which we are conscious is what can be seen during each fixation, and which is seen against a much less detailed world which is stable in world coordinates. Those working in the field tend to call this minimal world model '*gist*', although this somewhat dismissive term hardly does justice to its importance. I prefer the older terms 'focal' for the current, fovea-centred, view, and 'ambient' for the stable underlying sketch. To be of use, this *gist*, or ambient representation, needs to contain the identities of key objects and their locations, both within the current field of view and outside it. The need for such a mapping becomes very obvious when we have to locate objects during everyday tasks.

I would argue, then, that we do not make our conscious world picture by joining up fixations, even though we may retain a skeletal ambient world model also derived ultimately from visual input. Further evidence against a continuously updated pictorial model derived from fixations came from experiments in the 1990s, collectively referred to as change blindness. If the viewer is presented with a picture, and then, after a blank gap of about 80 milliseconds, the same picture is presented but with some major feature deleted or changed, that change will take a long time to detect (the loss of an engine from a plane is a common version). If, however, the pictures are presented with no gap—i.e. they are seen within the same fixation—the difference becomes instantly clear. Another version of this phenomenon is the 'spot-the-difference' puzzle often seen in magazines (Fig. 88). In one example, illustrated by Richard Gregory,[8] two men hold a conversation. It usually takes quite a long time long to spot the difference between the two pictures, and in this case, the only thing preventing this difference being detected is the saccade made from one picture to the other. Once again, if the two pictures are presented in the same place on a screen with no saccadic gap (I've done this in PowerPoint), the difference is immediately visible. Interestingly, if you *happen* to be looking at the right place in one picture before transferring gaze, the difference also becomes clear: attention circumvents the saccade-induced loss. Otherwise, it seems that each time a saccade is made, most of the pictorial content of the last fixation is lost. My colleague Ben Tatler did a simple experiment in which subjects were interrupted while engaged in a task (tea making) by turning out the light. He found that the subjects gave accurate and detailed accounts of what they saw just prior to the light going out, but they were unable to say anything useful about the fixation before that. Intriguingly, when the light went out just after a saccade, subjects

Fig. 88. A pair of 'spot the difference' pictures. Look from one picture to the other until the difference becomes clear.

could recall details of *either* the new fixation *or* the previous one, but not both: each new fixation overwrites the fixation before. Curiously, this over-writing could occur up to 0.4 seconds *after* the saccade, implying that the content of each fixation is maintained in some kind of very short-term memory buffer. The upshot of all these studies is that little survives a fixation change. This rules out the idea of an incremental model in which each fixation is added to the last: a kind of Bayeux tapestry where one scene follows on from the one before. The question then becomes what does survive? In my view this is the persisting ambient framework, which allows for navigation, object location, and gaze control.

In 2001 Kevin O'Regan and Alva Noë came up with what at the time was a quite radical scheme.[9] There were two main ideas. One was that there is no need for the visual system to construct a 'model of the world in the head'. In their view the brain does not have to form an explicit memory representation of the world: the world *is* our 'outside' memory. It is, after all, out there to be scrutinized. Why keep a dog and bark yourself? This seems sensible enough, but it doesn't address the question of where to look next. We might be prompted to move our eyes simply by the 'salience' of objects in the world, that is to say their eye-grabbing qualities such as size, brightness, contrast, detail, or motion. By and large this is not what happens: when engaged in a task, we look at objects and places that provide us with information and, generally, ignore other objects, however salient. Whilst agreeing with O'Regan and Noë that an internal model is not necessary for interrogating the world, I nevertheless believe that a stable ambient framework for planning future actions is essential. I will give my reasons in the section on 'vision and action'.

O'Regan and Noë's second idea was that seeing is a form of acting: a way of exploring the environment. Conscious experience of seeing occurs when the brain acts on this external representation to formulate words or actions: in their words, when it masters 'the governing laws of sensorimotor contingency'. This is a catch-all term for everything you have ever learnt about the ways that actions can be controlled by vision. A simple example would be looking at a mug and picking it up by the handle. This idea does away with the unnecessary internal pictorial model, but replaces it with something at least as complicated. I don't doubt that we have a repertoire of sensorimotor contingencies to deal with the ways vision is used to control actions, possibly including eye movements. My assumption is that these rules have been learnt, over the years, by the various Zeki-type nodes that make up our perceiving machinery, and in the corresponding motor control regions of the brain. However, I am less happy about attaching sensorimotor contingency to consciousness. When I look out of my window, there is the vivid scene, of which I am certainly conscious, without my having to act on it in any particular way. If I then examine some part of the scene, identifying a flower, noting a particular colour, or estimating the distance or speed of an approaching car, I could be said to be employing my knowledge of sensorimotor contingencies, and perhaps my consciousness is thereby heightened. But my intuition is that I have a conscious visual experience *anyway*, without a requirement for interaction, above what the perceptual machinery of the visual pathway automatically provides.

Vision and Action

Long before video eye-cameras were available, Alfred Yarbus, working in Russia in the 1950s, had shown that eye movements can be linked to particular tasks. He showed a painting by Repin, *The Unexpected Visitor*, to subjects whose eye movements he was monitoring with a photographic method.[10] The painting contained a woman, two children, a maid, and a man who has clearly just arrived from a journey (Fig. 89a). Yarbus asked the subjects different questions. 'Remember the clothes being worn': the eye movements are predominantly vertical following the dresses or coats of the people in the picture (Fig. 89c). My favourite is 'Estimate how long the unexpected visitor has been away from the family': the eye movements are predominantly horizontal, between the faces of the man, the woman, and the children (curiously not the maid), looking for clues to expressions such as surprise or joy (Fig. 89d). And similarly with other questions. What Yarbus had shown was that when there is a task to be done, the role of eye movements is to put the fovea in a position to find relevant information. Yarbus was not quite the first to point out this information-seeking role of eye movements (that distinction goes to Guy Buswell in the 1930s) but he did manage to demonstrate convincingly that eye movements are under 'top-down control', and are not just reflex responses to 'salient' objects that catch the eye.

In 1999 three of us—myself, Jenny Rusted, and Neil Mennie—wrote a paper on the gaze movements involved in making a cup of tea.[11] Many years after Yarbus, we were trying to see what his conclusions looked like when applied to a real task in real time, without external instructions from an experimenter. This seemingly mundane task involves 30–40 separate actions, involving many different objects and associated actions. The kettle has to be filled and monitored as it boils; mugs found and relocated; milk has to be

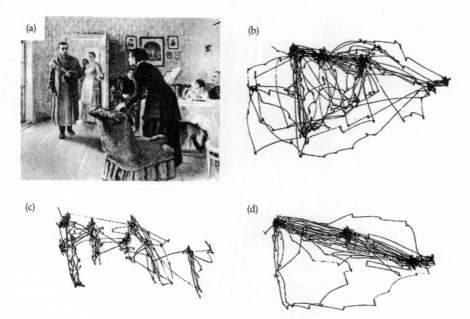

Fig. 89. Selection, from Alfred Yarbus,[10] of the eye movements made when viewing 'The Unexpected Visitor' by I.P. Repin. a) The picture. b) Free viewing for 3 minutes. c) Remember the clothes worn by the people. d) Estimate how long the visitor has been away from the family. Thin lines are saccades, and the interruptions and knots are fixations.

found and poured; sugar found; and so on. The eye-recording methods were as outlined in Chapter 5, but in addition to the eye camera, we had a second camera monitoring the tea-maker's actions (Fig. 90), so that we finished up with a quite comprehensive account of all the actions and the fixations that accompanied them. Unsurprisingly, the analysis was complicated, and it took about 6 months before we were sure what was going on. A key finding was that there was a predictable pattern. Each action involved a particular object and was preceded by a fixation on that object. This fixation occurred roughly half a second before the beginning of action on the object. So, in the examples in Figure 91, gaze first arrives on the kettle, which is scrutinized for several fixations. (It was a complicated electric kettle.)

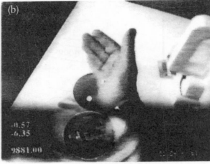

Fig. 90. Making a cup of tea. a) View from the scene-monitoring camera of the author, wearing the eye camera, putting a sweetener tablet into a mug. b) View from the eye camera at the same moment. The sweetener is still in the hand; gaze direction (white dot) is in the mug about to receive the tablet.

After three quick fixations to the sink, the kettle is picked up and gaze moves to the lid, which is taken off half a second later, while the kettle is being taken to the sink. The gaze then moves to the taps: three fixations—the right tap, the wrong tap, and the right tap again. Half a second later, the hand moves to the cold tap and turns it on. The gaze returns to the kettle, which is lowered to the sink, and gaze remains on the kettle until it becomes clear that it is full. Possibly you might have guessed all this, but several aspects of the sequence are less than obvious:

1. Eye movements are used to find the information the motor system needs, about half a second ahead of the action.
2. The eye-movement system knows where to go. The gaze hardly ever goes to an object that is not related to the task.
3. Towards the end of each action, gaze moves on to the next object in the sequence, again about half a second before the action is complete.
4. Vision is used to check that tasks are complete, for example, that the kettle has boiled, before the gaze can move on to the next object in the sequence.

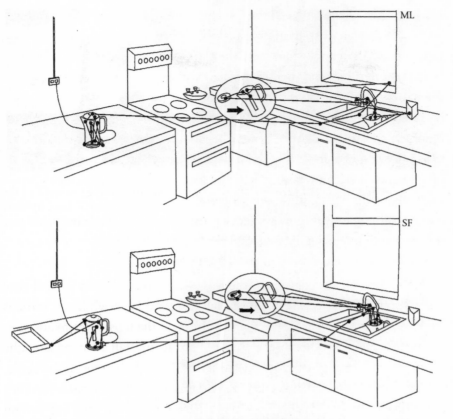

Fig. 91. Examples of the fixations (black dots) made by two participants filling a kettle from the sink. The fixation sequence is almost the same in the two cases. Nearly all fixations are on task-related objects. The two exceptions are a glance to the sink tidy on the right (upper record) and to a tray on the left (lower record).

In summary, gaze is used to locate and identify objects, to guide the hands to objects, to direct the hands when manipulating objects, and to check that each subtask is completed.

What impressed us, more than anything else, was the dedication of the eye-movement system to the task. With very few exceptions, the gaze sought only 'task-related objects'. There were plenty of other objects in the kitchen that could have 'caught the eye', but they hardly ever did. Mary

Hayhoe, then at the University of Rochester in the USA, got her students to do an equivalent transatlantic task of making peanut butter and jelly sandwiches. Her conclusions were the same: the eyes go to the next object in the required sequence before the hands start to move. In one version of the experiment, the table contained not just the objects required for the task—knife, bread, and plate—but many irrelevant objects such as pliers, hammer, and so on. Before the task began, the subjects made fixations to task-irrelevant objects 48% of the time. But once the task started this fell to 16%. Fixations became task specific as soon as the task became defined. Mary also commented that 'the sequences of fixations and reaches were almost identical for four subjects, despite the unspecific nature of the instructions'. It seems that different individuals develop convergent eye–hand strategies, at least in straightforward tasks. Figure 91 makes the same point.

These observations tell us a great deal about how sequences of actions are organized. Many parts of the brain are involved. First, there is the task itself. This sequence of actions has been learnt over a lifetime so that it is almost automatic and can be executed while listening to the news on the radio. Problems that occur in patients with dementia or other brain damage point to the frontal cortex and, in particular, the dorso-lateral prefrontal cortex (DLPFC), as the brain region where instruction sets for action sequences are stored (the 'schemas'). When the tea-making task is called for, the DLPFC has to issue instructions in the correct sequence to three other parts of the brain: the eye-movement system, the visual system, and the motor system. The job of the eye-movement system is to locate the specified object from the memory of its location, established during a brief preview of the kitchen, bearing in mind that the object may not be in the current field of view. When the object has been found, its location must be provided to the motor system. The visual system has to cooperate with the eye-movement system in identifying the object, and then cooperate with the motor system in shaping the response. And the motor system has to produce, from its own learnt repertoire, the appropriate behaviour. When the current subgoal has been achieved, the visual system has to determine that it is complete, and sign it off, so that the DLPFC can issue the next set of instructions. This routine must be repeated 30 or 40 times, but differently each time. This sounds complicated, and it is. Figure 92 is an attempt to represent the regions of the brain involved.

Apart from the DLPFC (the Schema control system in Fig. 92), the eye-movement system involves the frontal eye fields and parts of the parietal cortex; the visual system, as we have seen, involves the primary visual cortex and areas around it (Fig. 85) and the temporal lobe; and the motor system involves the parietal cortex, the premotor cortex, and the motor cortex itself, which issues the final commands to the limbs. This is before even considering the many regions beneath the cortex that are necessary for the

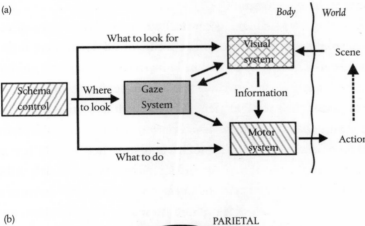

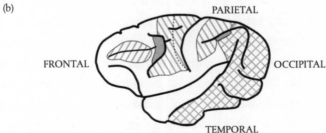

Fig. 92. a) The subsystems involved in the execution of an action in a sequence, indicating the interactions involved. b) The parts of the cerebral cortex physically involved in these actions.

operation of all these subsystems. Altogether this is something like half the brain, and all for making a cup of tea.

Clearly, the activation of a schema (e.g. 'fetch a mug') by the frontal cortex involves mechanisms for selecting particular components from the extensive repertoires of the three subsystems (gaze, vision, and motor) that execute the schema. The eye-movement system is ordered to direct gaze to the remembered location of a mug; the visual system is called upon to identify the mug; and the motor system is required to lift the mug from its hook and relocate it. These are all quite different from the requirements of any other schema, such as filling the kettle or pouring out the milk. How the selection systems operate is still a mystery, but it is clear that this must involve a great deal of neural traffic within and beyond the cortex.

The only selection mechanism that has been studied in any detail is the visual-attention system, and even then, the details of its working are far from clear. In 1980 Michael Posner used the expression 'the spotlight of attention' to refer to the way objects in the visual periphery can be singled out from the overall scene and in some sense 'highlighted'. In practical terms this means that objects or letters become easier to recognize in the attended region. The spotlight idea had its roots in William James' *Textbook of Psychology* (1890), where he writes: 'Everyone knows what attention is. It is the taking possession by the mind, in clear and vivid form, of one out of what seem several simultaneously possible objects or trains of thought.' More recently attention has become closely associated with the preparation of an eye movement. Two kinds of attention are recognized: *covert* attention, where an object in the periphery is singled out, but no eye movement is made, and *overt* attention, where an eye movement is made to the attended location. In practice it requires effort *not* to make an eye movement, once attention has been directed to a particular location.

In monkeys trained to attend to peripheral stimuli, recordings from cells in many visual regions of the cortex show evidence of attention. They have a higher firing rate and are more sensitive to local contrast. This influence extends even to cells of V1, at the input stage to the cortex. These enhancements are particularly striking in cells whose locations are about to become the targets of saccades. One suggestion is that this enhancement provides some advance information that will help the fovea to recognize that it has reached the right point, once the saccade is complete. Recent scanning studies (fMRI) suggest that enhancement can be quite specific: when subjects are cued to attend to colour or motion, the appropriate regions of visual cortex, V4 or V5, show increased responsiveness, implying that they have been specifically targeted by attention directed from 'higher' cortical regions.

In the visual subsystem of Figure 92, 'attention' seems to be the neural conduit from the frontal lobe that selects what should be looked for. But, as we have seen, during the execution of a task it is not just the visual system that is subject to selection, but so, in parallel, are the eye-movement system and the motor system. Should the word 'attention' be reserved solely for its role in sensory systems, or does it have a parallel function in selecting the appropriate components of all three subsystems? The exploration of these other control pathways is an important subject for future research.

Consciousness in Other Animals

If we accept Zeki's view, as I do, that consciousness arises in brain structures that analyse the attributes of the external world, such as colour, depth, and motion, then it seems reasonable to think that wherever analyses of similar complexity are carried out they too will be accompanied by conscious awareness. Among animals, that would include, at the very least, all vertebrates, arthropods, and the cephalopod molluscs. If one thinks, in evolutionary terms, of the possible benefits of consciousness, the one that stands out is that it provides an instantly accessible account of everything that is going on in the world: one that allows possibilities to be assessed and actions to be planned. Animals that have choices about what to do next are likely to need such an 'executive summary' of the state of the world, and a conscious image is a way of providing this.

It is easier to attribute consciousness to an animal that can demonstrate some of the behaviours that we would recognize as conscious in ourselves. With other mammals this is fairly easy, but it applies too to animals more distantly related to us. If you watch the caution with which a male jumping spider approaches a potentially hungry female, it is hard not to empathize with its predicament (Chapter 4). If you also know that it uses its eyes to actively scrutinize other animals for cues to their identity, and that it can plan detours to reach potential prey, then it becomes hard to resist the attribution of conscious awareness. Similarly with cephalopod molluscs, whose rapidly changing chromatophore patterns are used for communication, as well as camouflage, a cuttlefish can change its body pattern to blend in with a variety of substrates, but it can also produce flamboyant patterns, such as zebra stripes and eye spots, that are used to convey subtle meanings during courtship and in defence.[12] The rapidity of these changes implies an active, and presumably conscious, 'mind'.

In trying to apply our own, subjective, knowledge to other animals, we always seem to be stalked by zombies. In its philosophical form the zombie question is could a robot exist of such complexity that it is able to perform some or all lifelike functions competently? Perhaps self-driving cars are the closest things we have to this at present. Such 'autonomous' devices have currently no capabilities outside the roles they have been programmed to perform, but it is easy to imagine these roles being expanded to the extent that choices of various kinds become possible. If that were to happen, then one may have to contemplate the possibility that they too take on some aspects of consciousness: that they cease to be simply 'zombies'. We have to remember, too, that, as Dennett and others have pointed out, many of our own activities are carried out without conscious awareness. Walking or even driving requires little conscious involvement unless something unusual occurs. Similarly, the destination of our next eye movement is a decision usually made entirely without awareness. When thinking about other animals, even if we accept that behavioural complexity is a reasonable guide to conscious control, there is still no simple way of telling which aspects of their behaviour may require conscious oversight and which probably do not.

7

THE EVOLUTION OF VISION

This book has dealt with a wide range of topics, from eye diversity to human consciousness, but in this final chapter, I want to develop a theme they have in common. Towards the end of his life, in 1973, the great geneticist Theodosius Dobzhansky wrote a paper, 'Nothing in Biology Makes Sense except in the Light of Evolution'. Dobzhansky's paper was intended as an attack on creationism, but the idea is central to the way biologists approach the diversity of living things. If there is an advantage to be had by breathing air, flying, or seeing better, then Darwinian evolutionary mechanisms will find ways of making these things happen. This applies to the diversity of eyes—the subject of the first three chapters—but also to the ways that the images these eyes produce are exploited to provide an increasingly complex and versatile information base for the control of behaviour. This in turn has allowed those groups in which this process has gone furthest—the insects and the vertebrates—to diversify into almost all the habitats the world provides.

We have seen that eyes found in nature make use of nearly all the optical principles known in optical technology. These natural discoveries did not occur, or at least did not appear in the fossil record, until the Cambrian Period, starting about 540 million years ago. We know that photoreceptors had been present for much longer, because they are found, in recognizable form, in the cnidarians. This group, which includes the jellyfish and corals, was one of the first to emerge after multicellular animals came into existence, perhaps 800 million years ago. One might think that a photoreceptor on its own is of little use, but this would be wrong. Photoreceptors outside eyes can be used to establish daily rhythms, work out depth in the sea, respond to shadows, and, with a little screening pigment, work out light direction.

For several hundred million years in the Precambrian, these must have been what photoreceptors were used for. To make a basic eye, the first step would have been to put a number of receptors in a pit with a confined aperture, or into separate pigmented tubes. This would provide a modest amount of spatial resolution: enough, say, to avoid a rock in the path ahead or detect the movement of a large predator. 'Eyes' such as these would have been present at the dawn of the Cambrian. Then, within a few million years, compound eyes with many facets and good resolution appeared. A little later, we find fossils of fish-like creatures and cephalopod molluscs, and these animals would have had camera-type eyes with a single chamber and, presumably, lenses that produced decent images. The only kind of eye that we can be certain was *not* present by the end of the Cambrian was our kind of eye, where the main image-forming surface is the cornea that separates air from the fluid-filled eye cavity. That had to wait for the emergence of vertebrates onto land, about 360 million years ago, at the end of the Devonian Period.

There are only so many ways that light can be manipulated, and the laws of physics are the same whether the materials involved are glass or protein. Several times in evolution, different animal groups have arrived, independently, at the same optical solution to making an image. The most famous case of 'convergent evolution' is the extraordinary similarity of the eyes of the mollusc *Octopus* and the eyes of fish. In both cases the spherical lens is not optically homogeneous, but has a very particular refractive index profile which produces an excellent image free from the most important defect of lenses, spherical aberration. The focal length of such a lens is 2.5 times its radius, and this determines the shape of the rest of the eye, so that in external appearance the eyes are remarkably similar. It is only when the retinas are compared—*Octopus* receptors point towards the light, and fish receptors away from it—that it becomes clear that these eyes have separate ancestries. Eyes with this kind of lens also occur, sporadically, in crustaceans, gastropod molluscs, and annelid worms, and so evolved independently at least five times. Gradient lenses are not the only example of convergent optical evolution. The refracting superposition compound eye, which makes use of arrays of telescope-like lenses, produces a much brighter image than the more common apposition eye. It evolved in the insects (moths and beetles) to allow vision in near darkness, and independently in crustaceans (mysid shrimps and euphausiid krill) to make vision possible in the dim waters of the ocean.

Evolution is both conservative and inventive. Where a design works well, it tends to be retained, as in the case of the vertebrate eye. On the other hand, new designs arise from time to time. These include the mirror eyes of scallops and the even more remarkable mirror compound eyes of prawns and lobsters. Some eyes, one feels, should not still be there. The quite large pinhole eyes of *Nautilus* provide poor resolution, and yet have persisted through evolutionary time, even though *Nautilus'* relatives—squid, cuttle-fish, and *Octopus*—all have eyes with excellent lenses. The answer must be that eyes are metabolically expensive, and *Nautilus* is a grazer rather than a predator, so a high-resolution eye was never a priority. This is no doubt the reason why many animal groups, such as gastropod snails and annelid worms, have retained small, low-resolution eyes to supply their minimal visual needs at low cost. There is one family of gastropods (the heteropod sea snails) which have larger eyes and better resolution, and one family of worms (the alciopids) with eyes quite like those of small fish. It is significant that these animals are plankton-eating carnivores that need enhanced vision for predation. Evolution can make progress (as we might put it)—but only when there is a need for it.

A well-resolved image is, by itself, no guarantee that an animal will see the world in all its rich complexity. Scallops' eyes, with their mirror optics, have a resolution of about 2°, which is impressive for a bivalve mollusc. But their only visual behaviours consist of responding to movement, and pos-sibly crude steering when swimming. The compound eyes of some fan-worms resolve well, but again their only purpose is to detect predators. Seeing a moving predator at a distance *requires* an image-forming eye with good resolution, but whether any other features of the image can be dis-cerned depends on the capacity of an animal's brain to process information in multiple ways. This makes possible the multipurpose vision found only in the arthropods, the cephalopod molluscs, and the vertebrates. In all three groups, a particularly large proportion of the brain is devoted to vision.

If we look at the visual abilities of a 'top insect' such as a bee, these fall into two main categories: discerning the patterns of features in the world, and the patterns of motion caused either by moving objects in the world or by the bee's own movements. In the first category, bees learn the shapes and colours of the flowers they feed on, and the whereabouts of good feeding areas relative to the hive. They can even transmit this information to other bees via the famous dances described by Karl von Frisch. To find their way about bees use two kinds of spatial information: learning the patterns and sequences of landmarks between food and the hive, and using celestial cues, including the position of the sun and the pattern of polarization in skylight. These celestial cues are compensated for time of day by an internal clock. It seems unlikely that bees recognize each other as individuals, but in one species of paper wasp, *Polistes fuscatus*, the heads are marked with specific patterns which do indeed allow individuals to learn to recognize each other. Motion can signify the presence of danger, but its more common use is to monitor the bee's own movements. A key idea here is the 'flow-field'—the pattern of motion across the whole eye. This can be used to estimate velocity relative to the ground, to prevent deviations from a straight course, to navigate through confined spaces, and to make appropriate adjustments when coming in to land. This is a formidable list of behavioural competences, and one that is probably matched only in mammals or birds. It is interesting that bee vision is not particularly acute. Image resolution is about 100 times worse than our own, but that is still sufficient to permit a wide range of activities.

Bees are no doubt unusual in their well-developed ability to learn. Studies of insects and lower vertebrates have tended to concentrate on instincts, which involve making relatively stereotyped responses to specific stimuli. The dances of male jumping spiders when they detect a female—or rather, something that has the leg-pattern of another jumping spider—are a good example of a 'fixed-action pattern'. But even then the behaviour is 'fixed' only in its signalling components, which early ethologists called 'innate releaser mechanisms'. The male must adjust his dance so that it always faces the female, and adjust his leg movements to the terrain. The female does not just acquiesce. She may withdraw or, if hungry, attack the male. She is making decisions, even if learning is not involved. Programming key components of behaviour via the genes is one way to make sure that an animal with a small brain can respond fast and effectively to particularly important stimuli. The response of fiddler crabs to predators is a splendid example of how an innate behaviour can be simple but effective. Instead of trying to learn the characteristic features of particular predators, fiddler crabs use one rule: if it moves and protrudes above the horizon, it is likely to be a threat and requires an evasive response. Simplified signals are often involved in the recognition of one's own species, and are particularly important in courtship behaviour. In fiddler crabs it is not the physical appearance of the crab but the temporal patterns of the claw-waving displays that indicate species identity. In fireflies it is the pattern of flashing. In mantis shrimps it is a combination of colour and body polarization pattern. In crickets it is the song. In all these cases, the use of simple but specific badges of identity mean that the animals do not face the more complex task of working out identity from physical features that change their appearance whenever an animal moves.

Good vision, whether produced by compound or camera-type eyes, requires a still image on the retina. The reason is simply that photoreceptors are slow to respond because of the biochemical chain involved in producing an electrical response. A 10- or 20-millisecond response time may sound short, but translated into motion of the image across the retina it means that blur, like that with a moving camera, sets in at very low speeds. For humans, motion blur occurs at speeds above 1° per second: thus, gaze needs to be kept almost stationary. Convergent evolution has played a part here too. Vertebrates, insects, crustaceans, and cephalopod molluscs all show the same pattern of fixations, in which gaze is held stationary, separated by fast 'saccades' that relocate gaze direction. This alternation ensures that clear vision of the world takes place when the image is still and that relocating movements take up as little time as possible. Animals move, and so it is essential that during fixations, movements of the eyes are compensated for movements of the head and body. In all vertebrates, head rotations are monitored by the semi-circular canals of the inner ear; these send signals to the eye muscles, which move the eye by an amount equal and opposite to the head movement. In crabs this compensation is achieved using equivalent organs, the statocysts, which are quite similar to the vertebrate canals in structure, but have a completely separate evolutionary history. In two-winged flies, the rotation sensors are the halteres. These are modified second wings, short clubbed structures that vibrate vertically and behave as reciprocating gyroscopes. When the fly turns, these tend to hold their plane of vibration in space, and so stimulate receptors that effectively measure the fly's body rotation. Neck movements produce compensatory head movements that stabilize gaze. (The eyes are part of the head.) Just as eyes require images, so images require systems to keep them still for long enough to allow vision to occur. Evolution has repeatedly produced the same answer to the same problem, but by means that reflect different evolutionary histories.

It would be nice to end this book by having something useful to say about the evolution of visual consciousness. Clearly, we have no way to compare the subjective aspects of consciousness across species. Qualia (colour, timbre, and fragrance), which are so much a part of human conscious experience, remain impossible to pin down in humans, let alone other animals. If I cannot be certain that the blue I see is the same subjective colour as the blue you see, what chance is there that I can accurately imagine how a fish might see blue? We know that fish can see colour, but that is as far as we can go. Qualia seem destined to remain enigmatic.

There is one aspect of the problem where perhaps some progress can be made. If we knew what consciousness was good for, we could perhaps speculate as to which animals might need it. There would be reasons for it to evolve. Christof Koch, in *The Quest for Consciousness*, argues:

> The function of consciousness is to summarize the current state of the world in a compact representation and make this 'executive summary' accessible to the planning stages of the brain.... The content of this summary is the content of consciousness.

Many of our actions—eye movements being a good example—are executed by what Koch refers to as 'zombie agents'. These are automatic response mechanisms that do not intrude into consciousness. This will be as true in other animals as it is in man. The need for an executive summary arises when there are behavioural choices to be made, and information is needed on which to base these choices. It is probable that a snail in a pond has little choice most of the time, other than to consume palatable vegetation. Similarly, for much of its day, a male fiddler crab only has to feed sand into its mouth with its spare claw, respond in an automatic way to predators by evasion, and to passing females (or cotton reels!) by waving. On the other hand, the same crab is aware of the locations of its neighbours and their burrows: their activities generally produce no responses, hostile or otherwise. But if another male arrives and is not in a familiar location, he will be attacked. At the very least, fiddler crabs have a 'situational awareness' of what is going on around them, and this begins to look like Koch's executive summary. One could make a similar argument for any animal with a territory to defend; a robin will attack other males on its territory but not its mate. Using such examples, one could make a case for an evolutionary scale of consciousness, based on the complexities of the choices that different animals have to make.

Many animals have eyes to see. For some the uses of vision are limited, but for others 'seeing' means using visual information in many different ways. With the evolution of complex vision has come the need to choose between different behaviours. Such choices need to be managed by an 'oversight' system, and that would seem to involve conscious visual experience. We can only wonder, though, what it is like to experience the visual worlds of a honeybee or a mantis shrimp.

ENDNOTES

Chapter 1

1. Land, M.F. (1965). Image formation by a concave reflector in the eye of the scallop, *Pecten maximus*. *J. Physiol.* **179**, 138–53.
2. Action potentials are abrupt voltage changes that travel rapidly down nerves, and convey information in nervous systems.
3. Barnes, J. (editor) (1984). *The complete works of Aristotle.* Vol. I. p. 846. Princeton, NJ. Princeton University Press.
4. Nilsson, D.-E. (1998). Eyes as optical alarm signals in fan worms and ark clams. *Phil. Trans. Roy. Soc. B* **346**, 195–212.
5. Land, M.F. (2002). The spatial resolution of the pinhole eyes of giant clams (*Tridacna maxima*). *Proc. Roy. Soc. B* **270**, 185–8.
6. Eakin, R.M. (1975). *Great scientists speak again.* Berkeley and Los Angeles. University of California Press.
7. Arendt, D., Tessmar-Raible, K., Snyman, H., Dorresteijn, A.W., and Wittbrodt, J. (2004). Ciliary photoreceptors with a vertebrate-type opsin in an invertebrate brain. *Science* **306**, 869–71; Nilsson, D.-E. and Arendt, D. (2008). Eye evolution: the blurry beginning. *Current Biology* **18**, R1096–8.
8. Gould, S.J. (1989). *Wonderful life. The Burgess Shale and the nature of history.* London. Penguin Books.
9. Land, M.F. and Nilsson, D.-E. (2012). *Animal eyes* (2nd edn) Oxford. Oxford University Press.
10. Darwin, C. (1859). *The origin of species.* Chapter VI. London. John Murray.
11. Nilsson, D.-E. (2009). The evolution of eyes and visually guided behaviour. *Phil. Trans. Roy. Soc. B* **364**, 2833–47.

Chapter 2

1. Dick, O.L. (editor) (1949). *Aubrey's brief lives.* Penguin Classics Edition (1987). London. Penguin Books.
2. Larson, G. (1988). *The far side. Gallery 3.* p. 139. Kansas City. Andrews and McMeel.
3. Leeuwenhoek A. van (1695). In: *Arcana Naturae Detecta.* Delphis Batavorum. H. Kroonveld. Translated from the Latin by R. Wehner (1981).
4. Exner, S. (1891). *The physiology of the compound eyes of insects and crustaceans.* Original in German. Translated by R.C. Hardie (1989) and republished by Springer-Verlag.

5. Kirschfeld, K. (1976). The resolution of lens and compound eyes. In: *Neural principles in vision*. (eds. Zettler, F. and Weiler, R.) pp. 354–70. Berlin. Springer.

6. Collett, T.S. and Land, M.F. (1975). Visual control of flight behaviour in the hoverfly *Syritta pipiens* L. *J. Comp. Physiol.* **99**, 1–66.

7. Collett, T.S. and Land, M.F. (1978). How hoverflies compute interception courses. *J. Comp. Physiol.* **125**, 191–204.

8. Nilsson, D-E., Land, M.F., and Howard, J. (1988). Optics of the butterfly eye. *J. Comp. Physiol.* A **162**, 341–66.

Chapter 3

1. Denton, E.J. (1970). On the organization of reflecting surfaces in some marine animals. *Phil. Trans. Roy. Soc. B* **258**, 285–313.

2. Herring, P. (2002). *The biology of the deep ocean*. Chapter 9. Oxford. Oxford University Press.

3. Partridge, J.C. and Douglas, R.H. (1995). Far-red sensitivity of dragon fish. *Nature* **375**, 21.

4. Strictly speaking, the average used to blur the image is not a circle but a Gaussian function whose half-width is that of the circle shown.

5. Lockett, N.A. (1977). Adaptations to the deep-sea environment. In: *The visual system of vertebrates* (ed. Crescitelli, F.). *Handbook of sensory physiology* VII/5. Berlin. Springer, 62–192.

6. Wagner, H-J., Douglas, R.H., Frank, T.M., Roberts, N.W., and Partridge, J.C. (2009). *Dolichopteryx longipes*, a deep-sea fish with bipartite eyes using both refractive and reflective optics. *Current Biology* **19**, 108–14.

7. Land, M.F. (1981). Optics of the eyes of *Phronima* and other deep-sea amphipods. *J. Comp. Physiol.* **145**, 209–26.

8. Land, M.F. (1976). Superposition images are formed by reflection in the eyes of some oceanic decapod crustacea. *Nature* **263**, 764–5; Vogt, K. (1980). Die Spiegeloptik des Flusskrebsauges. The optical system of the crayfish eye. *J. Comp. Physiol.* **135**, 1–19.

9. Nilsson, D.-E. (1988). A new type of imaging optics in compound eyes. *Nature* **332**, 76–8.

10. Vukusic, P. and Sambles, R. (2003). Photonic structures in biology. *Nature* **424**, 852–5. Kinoshita, S. (2008) *Structural colors in the realm of nature*. Singapore. World Publishing Co.

11. The classical wave description of interference and reflection, given here, is not the one a contemporary physicist would use. Since the advent of quantum electrodynamics, introduced by Richard Feynman in 1985, in *QED* (Princeton University Press), it has been possible to describe refraction, reflection, scattering, and interference in terms of the ways that photons interact with each other. In bulk, photons still behave as classical waves, but this is a convenient illusion. A modern, accessible account is given by Johnsen, S. (2012) *The optics of life*. Princeton. Princeton University Press.

Chapter 4

1. Tinbergen, N. (1951). *The study of instinct*. Oxford. Oxford University Press. (reprint 1969).

2. Land, M. and Layne, J. (1995). The visual control of behaviour in fiddler crabs. I. Resolutions, thresholds and the role of the horizon. *J. Comp. Physiol.* A **177**, 81–90.

3. Salmon, M. and Atsaides, S.P. (1968). Visual and acoustical signalling during court-ship by fiddler crabs (genus *Uca*), *Amer. Zool.* **8**(3), 623–39.
4. Land, M.F. (1969). Structure of the retinae of the eyes of jumping spiders (Salticidae: Dendryphantinae) in relation to visual optics. *J. Exp. Biol.* **51**, 443–70.
5. Zurek, D.B., Cronin, T.W., Taylor, L.A., Byrne, K., Sullivan, M.G.L., and Morehouse, N.I. (2015). Spectral filtering enables trichromatic vision in colourful jumping spiders. *Current Biology* **25**, R403–4.
6. Nagata, T., Koyanagi, M., Tsukamoto, H., Saeki, S., Isono, K., Shichida, Y., Tokunaga, F., Kinoshita, M., Arikawa, K., and Terakita, A. (2012). Depth perception from defocus in a jumping spider. *Science* **335**, 469–71.
7. Land, M.F. (1969). Movements of the retinae of jumping spiders (Salticidae: Dendryphantinae) in response to visual stimuli. *J. Exp. Biol.* **51**, 471–93.
8. Harland, D.P. and Jackson, R.R. (2004). *Portia* perceptions: the *Umwelt* of an areneophagic jumping spider. In: *Complex worlds from simpler nervous systems* (ed. Prete, F.R.), pp. 5–40. Cambridge, MA. MIT Press. [*Areneophagic* means 'spider-eating'.]
9. Land, M.F., Marshall, N.J., Brownless, D., and Cronin, T. (1990), The eye movements of the mantis shrimp *Odontyllus scyllarus* Crustacea: Stomatopoda). *J. Comp. Physiol.* A. **167**, 155–66.
10. Marshall, N.J. and Oberwinkler, J. (1999). The colourful world of the mantis shrimp. *Nature (London)* **401**, 873–4.
11. Thoen, H.H., How, M.J., Chiou. T.-H., and Marshall, J. (2014). A different form of color vision in mantis shrimp. *Science* **343**, 411–13; Marshall, J. and Arikawa, K. (2014). Unconventional colour vision. *Current Biology* **24**, R1150–4.
12. Chiou, T.-H., Kleinlogel, S., Cronin, T., Caldwell, R., Loeffler, B., Goldizen, A., and Marshall, J. (2008). Circular polarization vision in a stomatopod crustacean. *Current Biology* **18**, 429–34.

Chapter 5

1. Carpenter, R.H.S. (1988). *Movements of the eyes*. London. Pion.
2. Walls, G.L. (1962). The evolutionary history of eye movements. *Vision Res.* **2**, 69–80.
3. Thomas, E.L. (1968). Movements of the eye. *Scient. Amer.* **219** (2), 88–95.
4. Land, M.F. (2009). *Looking and acting: vision and eye movements in natural behaviour.* Oxford. Oxford University Press.
5. Land, M.F. and Lee, D.N. (1994). Where we look when we steer. *Nature (London)* **364**, 742–4.
6. Buswell, G.T. (1920). An experimental study of the eye–voice span in reading. *Supplementary Educational Monographs* No. 17. Chicago. University of Chicago.
7. McConkie, G.W. and Rayner, K. (1975). The span of the effective stimulus during a fixation in reading. *Perception and Psychophysics* **17**, 578–86.7. Weaver, H.E. (1943). A study of the visual processes in reading differently constructed musical selections. *Psychological Monographs* **55**, 1–30.
8. Furneaux, S. and Land, M.F. (1999). The effects of skill on the eye–hand span during musical sight-reading. *Proc. Roy. Soc. B* **266**, 2435–40.

9. Regan, D. (1992). Visual judgments and misjudgements in cricket, and the art of flight. *Perception* **21**, 91–115.
10. Land, M.F. and McLeod, P. (2000). From eye movements to actions: how batsmen hit the ball. *Nature Neurosci.* **3**, 1340–5.
11. McKinley, T., Chajka, K., and Hayhoe, M. (2008). Pro-active gaze control in squash. *J. Vision* **8**(6), 111, 111a.

Chapter 6

1. Sacks, O. and Waterman, R. (1987). The case of the colorblind painter. *New York Review of Books.* Nov. 19.
2. Zeki, S. (2003). The disunity of consciousness. *Trends Cogn. Sci.* **7**, 214–18.
3. Dennett, D.C. (1991). *Consciousness explained.* New York. Little, Brown & Co.
4. Bridgeman, B. (2011). Visual stability. In: *The Oxford handbook of eye movements* (eds. Liversedge, S.P., Gilchrist, I.D., and Everling, S.), pp. 511–21. Oxford. Oxford University Press.
5. Robertson, C.E., Hermann, K.L., Mynick, A., Kravitz, D.W., and Kanwisher, N. (2016). Neural representations integrate the current field of view with the remembered 360° panorama in scene-selective cortex. *Current Biology* **26**, 2463–8.
6. Kim, H.R., Angelaki, D.E., and De Angelis, G.C. (2015). A functional link between MT neurones and depth perception based on motion parallax. *J. Neurosci.* **35**, 2766–77.
7. Wurtz, R.H. (2008). Neuronal mechanisms of visual stability. *Vision Res.* **48**, 2070–89.
8. Gregory, R.L. (1998). *Eye and brain* (5th edn). Oxford. Oxford University Press.
9. O'Regan, J.K. and Noë, A. (2001). A sensorimotor account of vision and visual consciousness. *Brain Behav. Sci.* **24**, 939–73.
10. Yarbus, A. (1967). *Eye movements and vision.* New York. Plenum.
11. Land, M.F., Mennie, N., and Rusted, J. (1999). The roles of vision and eye movements in the control of activities of daily living. *Perception* **28**, 1311–28.
12. Hanlon, R.T. and Messenger, J.B. (1996). *Cephalopod behaviour.* Cambridge. Cambridge University Press.

GLOSSARY

Action potential A brief electrical event that travels rapidly down a nerve. Action potentials result from a sudden exchange of sodium and potassium ions, and they are triggered by a change in voltage between the inside and outside of the nerve cell. This can be caused by sensory stimulation or by release of a chemical transmitter at a nerve junction (synapse).

Apposition eye A compound eye in which each lens is associated with a single receptive rhabdom—a rod composed of photopigment-bearing microvilli from seven to nine receptor cells. Each lens produces a small image at the outer tip of the rhabdom that defines the angle over which the rhabdom accepts light. There is no further resolution of the image within the rhabdom as light is scrambled by total internal reflection. Each lens-receptor complex is known as an ommatidium. The image seen by the animal is the erect image formed by the neural outputs of all the individual ommatidia (sometimes known as the mosaic image). Apposition eyes are typical of diurnal arthropods.

Bilateria Animals with bilateral symmetry, that are usually forwards moving and with a definite head end. Some early metazoans, notably the cnidarians (jellyfish and corals) have radial symmetry. All other phyla are bilaterians. Some bilaterians have become secondarily radial (e.g. starfish), but their larvae remain bilaterally symmetrical.

Bioluminescence Light production by living organisms. It is found in in bacteria, fungi, and animals and evolved independently many times. Bioluminescence works by the oxidation of a small pigment molecule, luciferin, catalyzed by an enzyme, luciferase, of which there are many types. The oxidation results in the emission of light.

Camera-type eye An eye with a single fluid-filled chamber and a single optical system consisting of a lens or, in terrestrial animals, a lens and cornea. The retina lines the rear of the chamber. The lens–gap–retina layout resembles the lens–gap–film (or sensor) layout of a typical camera. Camera-type eyes are typical of vertebrates, molluscs (especially cephalopods), and some spiders. They are sometimes also referred to as simple eyes, as opposed to compound eyes.

Chromatic aberration Image blur that occurs because blue light is refracted more strongly than red light, and, hence, lenses form blue images closer to the lens than red images, resulting in confusion and loss of resolution. There is no simple solution in nature, but having lenses with separate zones that refract light by different amounts can arrange for sharp blue and red images to be focussed in the same plane. However, the blue image remains contaminated by unfocussed red light, and vice versa, so this is only a partial solution.

Cilia Thin, finger-like structures containing a characteristic set of rodlets (microtubules), typically a ring of nine with two in the centre. Often cilia are motor structures which propel protozoans and move fluids through organs of the body. They also have the potential to develop into photoreceptors by having hugely expanded membranes containing photopigment. Human rods and cones are derived from cilia in which the membrane is folded into stacks of discs.

Compound eye A convex eye containing many separate lenses distributed across its surface. They are typical of insects and crustaceans.

Cornea The tough transparent layer at the front of the eye. In terrestrial vertebrates, such as ourselves, the cornea is part of the optical system, refracting light rays as they pass from air into the fluids of the eye.

Fixation Stationary periods of gaze between saccades, during which perception takes place. Because photoreception is slow, image motion during a fixation must be kept to a minimum to avoid blur. In humans, this stability is achieved by two reflexes: (1) the vestibular ocular reflex, in which the semi-circular canals of the inner ear detect movements of the head—this signal is used to move the eyes by an equal and opposite amount; (2) the opto-kinetic reflex, in which ganglion cells in the retina measure image movement, and this measurement is used to move the eyes in the same direction as the detected motion, thereby minimizing it. Other animals have equivalent stabilizing arrangements.

Fovea A region of the human retina close to the eye's axis with an especially high density of cones and no rods. Its diameter is about 1°—the size of a fingernail at arm's length. The fovea is the region of the eye which provides the highest resolution, and it is directed by saccades to objects of interest in the periphery or to places from which information is needed. The word *fovea* means 'pit'. The receptors are kept clear of the ganglion cells which overlie the rest of the retina. Many other vertebrates have foveas of various kinds, and compound eyes can have 'acute zones' whose functions are similar.

Ganglion A small group of neurones (nerve cells) and their connexions to each other or to sense organs or muscles.

Metazoa Multicellular animals, as opposed to the single-celled Protozoa.

Microvilli Narrow cylindrical projections from a cell body, typically less than 0.1 μm wide and 1 or more μm long (1 μm is one thousandth of a millimetre). In photoreceptors of many invertebrates, the membranes around the microvilli contain the photopigment molecules that receive light and produce an electrical change in the receptor cells.

Ommatidium A single element in an apposition compound eye consisting of the corneal lens and the receptive elements behind it. It contributes a single pixel to the overall apposition image.

Photophore A light-producing structure. These are very common on the bodies of marine animals from deeper water, where they can be used as attractors (anglerfish), illuminators (headlamp fish), defence (some crustaceans), and most commonly, as camouflage to disguise an animal's silhouette against light from the surface. In addition to light-producing material (see *bioluminescence*) photophores may have lenses, reflectors, and filters to ensure a good match to the colour and distribution of light from above. On land photophores occur as attractors in fireflies and in some other insects.

Photopigment A molecule, such as rhodopsin, that responds to light. In the unbleached state rhodopsin is purple, but it loses this colour when exposed to light.

Photoreceptor The receptors responsible for absorbing and responding to light. The molecules involved are photopigments, such as rhodopsin, which change their shape when they absorb single photons of light. This sets off a series of biochemical reactions which results in a change in voltage at the cell membrane. This in turn results in either transmitter release at a synapse or an action potential, depending on the type of eye.

Polarization of light Light reaches the eye as photons, packets of energy in the form of electromagnetic waves. These waves have an electrical and a magnetic field vibrating at right angles to each other. By convention, the plane of the electrical vibration (the E-vector) is referred to as the plane of polarization. Direct light from the sun has photons with E-vectors in all possible planes, and so is unpolarized. However, scattering of light by the atmosphere selects for photons with specific E-vector directions, and so light from the sky becomes polarized. This property is used, especially by insects such as bees, to determine the position of the sun in overcast conditions. Reflecting surfaces, such as water, also polarized light, and aquatic insects make use of this. Detecting polarized light requires photopigment molecules that are aligned with the E-vector of the photons. Invertebrate microvillous receptors provide such an alignment, but the discs of vertebrate receptors do not, which is why humans are unable to detect the plane of polarized light.

Protostome/deuterostome Ancient divisions of the animal kingdom based on embryology. The terms are Greek for 'first mouth' and 'second mouth', and they refer to the different ways that the mouth and anus develop in the early embryo. The two groups separated from each other in the Precambrian. The protostomes comprise the arthropods, molluscs, annelid worms, nematodes, flatworms, and several smaller phyla. The deuterostomes include the echinoderms (starfish relatives) and the chordates, which gave rise to the vertebrates and ultimately ourselves.

Pseudopupil The small, dark spot in a compound eye that appears to move around the eye as the observer views the eye from different directions. It appears dark because it absorbs light from the direction of the observer's eye, and so defines the direction of sight of that part of the eye surface. It is not a true pupil like that in a vertebrate eye.

Qualia Subjective qualities, such as perceived colour, sound, and smell that are personal to the observer and cannot be assessed by external measurement. Qualia provoke unanswerable questions such as how do I know whether red looks the same to you as it does to me? Qualia keep philosophers awake at night.

Refractive index A measure of a material's ability to bend light. It arises because the speed of light is faster in some materials than in others. At a surface separating two materials with refractive indices n_1 and n_2, the amount by which a ray of light is bent is given by Snell's law: $n_1 \sin i_1 = n_2 \sin i_2$, where i_1 and i_2 are the angles in the first and second media between the ray and a right angle to the surface. The refractive indices of air, water, tissue fluid, and dry protein are 1, 1.33, 1.34, and 1.52, respectively. Glass types vary from 1.5 to 1.75.

Retina The layer at the back of the eye containing the light receptors onto which the image falls. In humans these receptors are either rods, responsible for dim-light vision, or cones of three types that provide colour vision in brighter conditions.

Rhabdom Photoreceptor structures found in insects and other invertebrates. They are made up of large numbers of microvilli (thin, finger-like processes) whose membranes hold the photopigment. In insects and crustaceans, each rhabdom is formed of contributions from seven to nine receptor cells, fused into a single rod. *Rhabdom* is Greek for 'rod'.

Rhabdomere The stack of microvilli that make up the contribution of a single receptor to a rhabdm. In two-winged flies (Diptera), the rhabdomeres do not join to form a rhabdom but remain separate and receive different parts of each image. These contribute to a *neural superposition* image, explained in Chapter 2.

Rhodopsin A photopigment consisting of a protein (opsin) enclosing a molecule known as a chromophore, in most cases a derivative of Vitamin A. This is a long molecule with alternating double and single carbon bonds, and when it absorbs a photon of light, one of these bonds (number eleven) flips from a bent to a straight configuration. This sets off a biochemical chain reaction which results in a voltage change in the cell membrane, and starts the process of vision. The structure of the opsin protein surrounding the chromophore affects the range of wavelengths to which the chromophore responds. Our three cone types have the same chromophore but different opsins; they respond to different regions of the spectrum and make colour vision possible.

Saccade A rapid movement of eye or head or both which changes the direction of view of an eye. Human saccades are made about three times a second. Each lasts between 20 and 80 milliseconds, depending on size, and may reach speeds of 500 degrees per second. Humans are effectively blind during saccades. Insects also make saccades, but they turn their heads since their eyes are fixed in the head.

Spherical aberration Image blur, resulting from the fact that rays entering the outer zones of a lens form images closer to the lens than rays entering the inner zones of the lens near its axis. This produces image confusion and poor resolution. It can be counteracted by having a lens with a variable refractive index, so that its outer regions refract light less than in a homogeneous lens (fish lenses). Another solution is to have a lens surface that is not spherical (the human cornea).

Superposition eye A compound eye in which the contributions of the optical elements are combined (superposed) to produce a single, erect, deep-lying image. This combined image is much brighter than the individual images in apposition eyes, and superposition eyes are found in nocturnal insects (moths and beetles) and deep-water crustaceans. The optical elements are not simple lenses, but they function as two-lens telescopes, which invert the direction of entering light rays (refracting superposition). An alternative, found in shrimps and lobsters, is an array of mirrors forming square boxes at right angles to the eye surface (reflecting superposition).

Total internal reflection Light rays inside a high-refractive-index cylindrical structure, such as a rhabdom, become trapped if the angle they make with the cylinder's surface is too large. This is a consequence of Snell's law (see *refractive index*), and it occurs when angle i_1 (inside the cylinder) is sufficiently large that i_2 (the exit angle) would be greater than 90°. This occurs when $\sin i_1 > n_2/n_1$. For a rhabdom with a refractive index of 1.37, this would mean any ray making an angle of more than about 12° with the rhabdom wall $(90° - n_1)$ will become trapped inside.

Visual cortex The region at the back of the cerebral cortex of the mammalian brain (the occipital lobe) that receives the input from the eyes, relayed by the lateral geniculate nucleus of the thalamus, but, otherwise, little changed from the output of the retinal ganglion cells. Each half-cortex receives its input from one-half of the right field of view from both eyes: the left field goes to the right cortex and vice versa. The main function of the primary visual cortex (also known as striate cortex, from its anatomical appearance, or simply V1) is to extract lines and edges from the input it receives. This provides a cartoon-like representation of the image, which is then augmented by other cortical areas, responsible for motion, colour, and depth. It is far from clear how these multiple representations produce our single conscious view of the world.

ART CREDITS

Chapter 1

Figure 1 Michael Land, *The Eye VSI* (Oxford University Press, 2014) Figure 7c.

Figure 2a Michael Land, Figure 6.2b in Michael Land, *Animal Eyes*, 2nd edn (Oxford University Press, 2012).

Figure 2b Michael Land, Figure 6.3, left, in Michael Land, *Animal Eyes*, 2nd edn (Oxford University Press, 2012).

Figure 5a Nhobgood Nick Hobgood, CC-BY-SA-3.0.

Figure 5b Michael Land.

Figure 5c Michael Land, *The Eye VSI* (Oxford University Press, 2014) Figure 7d, photo courtesy of Dan-Eric Nilsson.

Figure 7a R. M. Eakin (1965). Evolution of Photoreceptors, *Cold Spring Harbor Symposia on Quantitative Biology* V.30, 364, Figure 1. Copyright holder, Cold Spring Harbor Laboratory Press.

Figure 7b Photo Professor Richard Eakin.

Figure 8 Based on Figure 1.7 in Michael Land, *Animal Eyes*, 2nd edn (Oxford University Press, 2012).

Figure 10a Figure 1.5 in Michael Land, *Animal Eyes*, 2nd edn (Oxford University Press, 2012).

Figure 10b Photo courtesy of Dan-Eric Nilsson.

Figure 11 Based on Figure 5.9 in Warrant and Nilsson (eds), *Invertebrate Vision* (Cambridge University Press, 2006).

Figure 12a/b/c Figure 4.5 in Michael Land, *Animal Eyes*, 2nd edn (Oxford University Press, 2012).

Chapter 2

Figure 13 Robert Hooke, *Micrographia* (1664).

Figure 14 Michael Land.

Figure 15 Figure 8.2, left, in Michael Land, *Animal Eyes*, 2nd edn (Oxford University Press, 2012).

Figure 16 Image courtesy of Kuno Kirschfeld.

Figure 17 Michael Land.

Figure 18 T. S. Collett and Michael Land (1975). Visual Control of Flight Behaviour in the Hoverfly *Syritta pipiens, J. Comp. Physiol. A,* **99,** 1–66, Figure 21.

Figure 19 T. S. Collett and Michael Land. Visual Control of Flight Behaviour in the Hoverfly *Syritta pipiens. J. of Comp. Physiol.* A, 99, 1–86 (Springer, 1975) Fig. 32A.

Figure 21b Figure 62 from H. Grenacher, *Untersuchungen über das Sehen der Arthropoden insbesondere der Spinnen, Insecten und Crustaceen. Gttingen* (Vanderhoeck & Ruprecht, 1879).

Figure 21c Michael Land.

Figure 22 Michael Land.

Figure 24 Compiled from Figures 7.16 and 7.17 in Michael Land, *Animal Eyes,* 2nd edn. (Oxford University Press, 2012).

Figure 25 Michael Land. Figure 8.2, right, in Michael Land, *Animal Eyes,* 2nd edn. (Oxford University Press, 2012).

Figure 26a Modified from S. Exner, *Die Physiologie der facettierten Augen von Krebsen und Insekten* (Deuticke, Leipzig, 1891) Plate 1, Figure 1.

Figure 26b Modified from S. Exner, *Die Physiologie der facettierten Augen von Krebsen und Insekten* (Deuticke, Leipzig, 1891) text, Figure 11.

Figure 27a Modified from S. Exner, *Die Physiologie der facettierten Augen von Krebsen und Insekten* (Deuticke, Leipzig, 1891) text, Figure 5.

Figure 27b Modified from S. Exner, *Die Physiologie der facettierten Augen von Krebsen und Insekten* (Deuticke, Leipzig, 1891) text, Figure 7.

Figure 27c Figure 8.4 in Michael Land, *Animal Eyes,* 2nd edn (Oxford University Press, 2012).

Figure 28a Justin Marshall.

Figure 28b Figure 8.9 in Michael Land, *Animal Eyes,* 2nd edn (Oxford University Press, 2012).

Figure 29b Figure 8.12a in Michael Land, *Animal Eyes,* 2nd edn (Oxford University Press, 2012).

Figure 29c Figure 8.12b in Michael Land, *Animal Eyes,* 2nd edn (Oxford University Press, 2012).

Figure 30 Figure 8.11 in Michael Land, *Animal Eyes,* 2nd edn (Oxford University Press, 2012).

Figure 31a Michael Land.

Figure 31b Michael Land.

Chapter 3

Figure 32 Howard Roe/National Oceanography Centre.

Figure 33 Michael Land.

Figure 34 Adapted from Figures 6 and 7 in E. J. Denton (1970). On the Organization of Reflecting Surfaces in Some Marine Animals, *Phil. Trans. R. Soc. Lond. B*, 258, 285–313, by permission of the Royal Society.

Figure 35 Justin Marshall.

Figure 36 a) Figure XI,6 from N. B. Marshall, *Aspects of Deep Sea Biology* (Hutchinson, 1954). b) Spothead lanternfish, *Diaphus metopoclampus*. From Plate 27 of *Oceanic Ichthyology* by G. Brown Goode and Tarleton H. Bean (1896). Public Domain, PD-US.

Figure 37 Figure XI, 10 from N. B. Marshall, *Aspects of Deep Sea Biology*, (Hutchinson, 1954).

Figure 38 Michael Land.

Figure 39 Figure 4.11 in Michael Land, *Animal Eyes*, 2nd edn (Oxford University Press, 2012); b and c modified from N. A. Lockett, Adaptations to the Deep-Sea Environment, in F. Crescitelli (ed.), *Handbook of Sensory Physiology*, Vol. VII/5 (Springer, 1977) pp. 67–192 after figures by Ole Munk.

Figure 40a Ron Douglas.

Figure 40b + c Figure 6.5 in Michael Land, *Animal Eyes*, 2nd edn (Oxford University Press, 2012). Based on H.-J. Wagner, R. H. Douglas, T. M. Frank, N. W. Roberts, and J. C. Partridge (2009). *Dolichopteryx longipes*, a Deep-Sea Fish with a Bipartite Eye Using Both Refractive and Reflective Optics, *Current Biology* 19, 108–14 (2009) Figure 1.

Figure 41a Figure 7.19a in Michael Land, *Animal Eyes*, 2nd edn (Oxford University Press, 2012).

Figure 41b Michael Land.

Figure 41c Figure 7.19a in Michael Land, *Animal Eyes*, 2nd edn (Oxford University Press, 2012).

Figure 42 Figure 7.19a in Michael Land, *Animal Eyes*, 2nd edn (Oxford University Press, 2012).

Figure 43a Figure 8.10 in Michael Land, *Animal Eyes*, 2nd edn (Oxford University Press, 2012).

Figure 43b N. B. Marshall. *Developments in Deep-Sea Biology* (Blandford Press, 1979) Figure 140, p. 399.

Figure 44 Michael Land.

Figure 46a Figure 8.13 in Michael Land, *Animal Eyes*, 2nd edn (Oxford University Press, 2012).

Figure 46b Figure 8.13 in Michael Land, *Animal Eyes*, 2nd edn (Oxford University Press, 2012).

Figure 46c H. Grenacher, *Untersuchungen über das Sehorgan der Arthropoden: ins besondere der Spinnen, Insecten und Crustaceen* (Vandenhoek & Ruprecht, Göttingen, 1879) Figure 119.

Figure 47 H. Grenacher, *Untersuchungen über das Sehorgan der Arthropoden: ins besondere der Spinnen, Insecten und Crustaceen* (Vandenhoek & Ruprecht, Göttingen, 1879) Figure 119.

Figure 48 Modified from Figure 8.15 in Michael Land, *Animal Eyes*, 2nd edn (Oxford University Press, 2012).

Figure 49 Modified from Figure 8.16 in Michael Land, *Animal Eyes*, 2nd edn (Oxford University Press, 2012).

Figure 50 Figure 6.12 in Michael Land, *Animal Eyes*, 2nd edn (Oxford University Press, 2012).

Figure 51 Michael Land.

Figure 52a + b Originally published in Michael Land (1972). The Physics and Biology of Animal Reflectors, *Progress in Biophysics & Molecular Biology* 24, 77–106, Figure 13.

Figure 52c Figure 6.17d in Michael Land, *Animal Eyes*, 2nd edn (Oxford University Press, 2012), adapted from Figure 8 in E. J. Denton and J. A. C. Nicol (1965). Reflexion of Light by External Surfaces of the Herring *Clupea harengus*,

J. Mar. Biol. Assoc. U.K. 45, 711–38, reproduced with permission.

Chapter 4

Figure 53 Figure 5 from Michael Land and John Layne (1995). The Visual Control of Behaviour in Fiddler Crabs. I. Resolution, Thresholds and the Role of the Horizon, *J. Comp. Physiol. A* 177, 81–900.

Figure 54 Figure 7 from Michael Land and John Layne (1995). The Visual Control of Behaviour in Fiddler Crabs, *J. Comp. Physiol. A* 177, 81–900.

Figure 55 Figure 7 from M. Salmon and S. P. Atsaides (1968). Visual and Acoustical Signalling during Courtship by Fiddler Crabs (Genus Uca), *American Zoologist* 8(3), 623–39, copyright Oxford University Press.

Figure 56 Michael Land.

Figure 57 Figure 12 from M. F. Land (1969b). Movements of the Retinae of Jumping Spiders (Salticidae: Dendryphantinae) in Response to Visual Stimuli, *J. Exp. Biol.* 51, 471–93.

Figure 58a Michael Land.

Figure 58b Modified from Figure 5.19 in Michael Land, *Animal Eyes*, 2nd edn (Oxford University Press, 2012).

Figure 59 Figure 5.19 in Michael Land, *Animal Eyes*, 2nd edn (Oxford University Press, 2012), originally, Figures 3 and 5 in Michael Land (1969). Structure of the Retinae of the Eyes of Jumping Spiders (Salticidae: Dendryphantinae) in Relation to Visual Optics, *J. Exp. Biol.* 51, 443–70.

Figure 60 Modified from Figure 1 from Michael Land (1969). Movements of the Retinae of Jumping Spiders (Salticidae: Dendryphantinae) in Response to Visual Stimuli, *J. Exp. Biol.* 51, 471–93.

Figure 61a Plate 1 from M. F. Land (1969). Movements of the Retinae of Jumping Spiders (Salticidae: Dendryphantinae) in Response to Visual Stimuli, *J. Exp. Biol.* **51**, 471–93.

Figure 61b Figure 9.13 in Michael Land, *Animal Eyes*, 2nd edn (Oxford University Press, 2012), originally, from M. F. Land (1969). Movements of the Retinae of Jumping Spiders (Salticidae: Dendryphantinae) in Response to Visual Stimuli, *J. Exp. Biol.* **51**, 471–93.

Figure 62a G. B. Edwards and David E. Hill (2008). Representatives of the North American Salticid Fauna, Revisited. *Peckhamia* **30**(2), 1–15. Photographed by G. B. Edwards/CC-BY-SA-3.0.

Figure 62b Michael Land.

Figure 63 R. A. Lydekker. *The Royal Natural History Vol 6* (1896). Public Domain.

Figure 64 Author's own research images. Originally appeared as Figure 2 from M. F. Land, N. J. Marshall, D. Brownless, and T. Cronin (1990). The Eye Movements of the Mantis Shrimp *Odontodactylus scyllarus* (Crustacea: Stomatopoda), *J. Comp. Physiol. A* **167**, 155–66.

Figure 65 Reprinted from *Current Biology* 24, Justin Marshall and Kentaro Arikawa, Unconventional Colour Vision, pp. R1150–4, Copyright (2014), with permission from Elsevier.

Figure 67 Figure 2.9 in Michael Land, *Animal Eyes*, 2nd edn (Oxford University Press, 2012).

Figure 68 Figure 2.9 in Michael Land, *Animal Eyes*, 2nd edn (Oxford University Press, 2012).

Figure 69a Figure 2.10 in Michael Land, *Animal Eyes*, 2nd edn (Oxford University Press, 2012).

Figure 69b Figure 2.10 in Michael Land, *Animal Eyes*, 2nd edn (Oxford University Press, 2012).

Figure 69c Reprinted from *Current Biology* 18, Tsyr-Huei Chiou, Sonja Kleinlogel,Tom Cronin, Roy Caldwell, Birte Loeffler, Afsheen Siddiqi, Alan Goldizen, and Justin Marshall, *Circular Polarization Vision in a Stomatopod Crustacean*, pp. 429–34, copyright (2008) with permission from Elsevier.

Figure 70 Ben Vincent.

Chapter 5

Figure 71 Michael Land and Benjamin F. Tatler, *Looking and Acting* (Oxford University Press, 2009) Figure 2.7.

Figure 72 Michael Land and Benjamin F. Tatler, *Looking and Acting* (Oxford University Press, 2009) Figure 1.4.

Figure 73 Michael Land and Benjamin F. Tatler, *Looking and Acting* (Oxford University Press, 2009) Figure 7.3, modified from Figure 8.2 in Michael Land, The Visual Control of Steering, in Laurence Harris and Michael Jenkin (eds), *Vision and Action* (Cambridge University Press, 1998).

Figure 74 Michael Land and Benjamin F. Tatler, *Looking and Acting* (Oxford University Press, 2009) Figure 7.5, modified from M. Land and D. N. Lee (1994). Where We Look When We Steer, *Nature* 369, 742, Figure 1.

Figure 75 Michael Land and Benjamin F. Tatler, *Looking and Acting* (Oxford University Press, 2009), Figure 7.6, modified from Figure 8.4a in Michael Land, The Visual Control of Steering, in Laurence Harris and Michael Jenkin (eds) *Vision and Action* (Cambridge University Press, 1998).

Figure 76 Michael Land and Benjamin F. Tatler, *Looking and Acting* (Oxford University Press, 2009) Figure 7.10, modified from Figure 8.9b in Michael Land, The Visual Control of Steering, in Laurence Harris and Michael Jenkin (eds), *Vision and Action* (Cambridge University Press, 1998).

Figure 77 G. T. Buswell, *An Experimental Study of the Eye–Voice Span in Reading, Supplementary Educational Monographs No. 17* (Chicago University Press, 1920) Plate 4.

Figure 78a + b Michael Land and Benjamin F. Tatler, *Looking and Acting* (Oxford University Press, 2009) Figure 4.6, modified from H. E. Weaver, *A Study of Visual Processes in Reading Differently Constructed Musical Selections, Psychological Monographs 55* (1942), Figure 2.

Figure 78c Figure 2 from Furneaux and Land (1999). The Effects of Skill on the Eye-Hand Span during Musical Sight Reading, *Proc. Roy. Soc. B* 266, 2435–40.

Figure 79 Michael Land and Benjamin F. Tatler. *Looking and Acting* (Oxford University Press, 2009) Figure 8.8, original from Beldan and Fry, *Great Batsmen, Their Methods at a Glance*, (Macmillan, 1905) out of copyright. Modified following D. Regan (1992). Visual Judgements and Misjudgements in Cricket, and the Art of Flight, *Perception* 21, 91–115.

Figure 80 Michael Land and Benjamin F. Tatler (2000). *Looking and Acting* (Oxford University Press, 2009) Figure 8.11, modified from M. F. Land and P. McLeod (2000). From Eye Movements to Actions: How Batsmen Hit the Ball, *Nature Neuroscience* 3, 1340–5, Figure 1.

Figure 81a + b Michael Land and Benjamin F. Tatler (2000). *Looking and Acting* (Oxford University Press, 2009) Figure 8.9, modified from M. F. Land and P. McLeod (2000). From Eye Movements to Actions: How Batsmen Hit the Ball, *Nature Neuroscience* 3, 1340–5, Figures 2 and 3.

Figure 82 Michael Land and Benjamin F. Tatler (2000). *Looking and Acting* (Oxford University Press, 2009) Figure 7.10, originally M. F. Land and P. McLeod (2000). From Eye Movements to Actions: How Batsmen Hit the Ball, *Nature Neuroscience* 3, 1340–5, Figure 5a.

Figure 83 Michael Land and Benjamin F. Tatler (2000). *Looking and Acting* (Oxford University Press, 2009) Figure 8.7, modified from T. McKinney, K. Chajka, and M. Hayhoe (2008). Pro-Active Gaze Control in Squash, *Journal of Vision* 8(6), 111, 111a, with added information.

Figure 84 Modified from Michael Land, *The Eye VSI* (Oxford University Press, 2014) Figure 34.

Chapter 6

Figure **86** Michael Land, *The Eye VSI* (Oxford University Press, 2014) Figure 25.

Figure **87** Michael Land, *The Eye VSI* (Oxford University Press, 2014) Figure 26, after K. Brodmann (1914) Physiologie der Gehirne, in *Die Allgemeine Chirurgie der irnkrankheiten, Neue Deutsche Chirurgie, Vol. 11: Enke*. From J. P. Frisby, *Seeing* (Oxford University Press, 1979), p. 79.

Figure **88** Figure 3.17 in Richard L. Gregory, *Eye and Brain* (Oxford University Press, 1998), copyright Oxford University Press.

Figure **89** Michael Land, *The Eye VSI* (Oxford University Press, 2014) Figure 22, reprinted/adapted by permission from A. Yarbus, *Eye Movements during Perception of Complex Objects, Eye Movements and Vision* (Springer Nature: Springer, 1967) pp. 171–77. Copyright Springer Science+Business Media New York (1967).

Figure **90** Michael Land and Benjamin F. Tatler, *Looking and Acting* (Oxford University Press, 2009) Figure 5.2, Michael Land, Neil Mennie, and Jennifer Rusted (1999). The Roles of Vision and Eye Movements in the Control of Activities of Daily Living, *Perception* 28(11), 1311–28.

Figure **91** Michael Land and Benjamin F. Tatler, *Looking and Acting* (Oxford University Press, 2009) Figure 5.2, modified from Michael Land, Neil Mennie, and Jennifer Rusted (1999). The Roles of Vision and Eye Movements in the Control of Activities of Daily Living, *Perception* 28(11), 1311–28.

Figure **92** Michael Land and Benjamin F. Tatler, *Looking and Acting* (Oxford University Press, 2009) Figure 1.1.

Plate credits

Plate **1** Matthew Krummins/CC BY 2.0.

Plate **2** Thomas Shahan/CC BY-NC-ND 2.0.

Plate **3** Ivan Kuzmin/Shutterstock.com.

Plate **4** annop youngrot/Shutterstock.com.

Plate **5** © Jurgen Otto.

Plate **6** fenkieandreas/Shutterstock.com.

Plate **7** Courtesy of Justin Marshall, the University of Queensland.

Plate **8** Public domain.

INDEX

action potential 183
acute zone 40, 114
afocal apposition 57
Airy disc 33
ambient vision 159–60
Angel, Roger 84
angler fish 66
Anomalocaris (Cambrian arthropod) 16
anticipatory eye movements 145–6
apposition eyes 18, 28, 32, 43, 74, 114, 183
Arendt, Detlev 14
Aristotle 4
Atsaides, Samuel 97
attention 168–9
   covert 168
   overt 168
Aubrey, John 28

Barbatia (clam) 8
Barber, Vernon 6
Bathylychnops (deep-sea fish) 72
batsman 142
Beaufort, NC 93
bees 174
Berkeley, CA 10, 100
Bilateria 13, 183
bivalve molluscs 8
bioluminescence 65, 183
blowfly 46
bounce point 142–3
Bridgeman, Bruce 152
Buswell, Guy 137, 162
butterfly eye 55

Caldwell, Roy, 112
Cambrian

explosion 15
   period 7, 13, 130, 171
camera-type eye 16, 24, 184
camouflage 64–5
Caneletto 155–6
cephalopod molluscs 170
cerebral achromatopsia 150
change blindness 159
Chiou, Tsyr-Huei 118
chromatic aberration 183
chunking 140
ciliary receptor 6, 14, 184
cnidarians 14
Collett, Tom 35
colour
   filter 104, 115–17
   vision, 149
compound eyes 8, 16, 184
concave mirror 3
cones 149
consciousness 148–9, 151, 169, 176
constructive interference 86
convergent evolution 172
cornea 184
corner reflector 81
counter-illumination 66, 68–9
cricket 140–5
Cystisoma (amphipod) 76

Darwin, Charles 18
Dennett, Daniel 152, 170
Denton, Eric, 3, 64, 88, 91
depth vision 154
Descartes, René 152
deuterostomes 11, 185
DeVoe, Robert 103

Devonian period 172
diffraction 33, 60
Dingle, Hugh 112
Diptera 40
Discovery (research ship) 61–2
disparity 154
distance estimation 105, 114
Dobzhansky, Theodosius 171
Dolichopteryx (deep-sea fish) 72–3
dorsal light reflex 64
Douglas, Ron 72
dragon fish 66
dragonflies 40, 46
Drees, Oscar 98
driving 132

Eakin, Richard 10, 101
efference copy 158
euphausiids 77
E-vector 119–20
evolution 171–7
executive summary 176
Exner, Sigmund 31, 47, 81, 114
eye camera 130–3, 137
eye-glow 52
eye movements
    in ball sports 140–7
    in driving 132–7
    in reading 137–8
    in reading music 138–40
eye movement system 167
eye–note span 140
eye–voice span 137

fiddler crab 46, 93–8
    courtship 84, 97
    eyes 95
    eye movement 107–8
    predator response 95–6, 175
fixation 129, 160, 164, 165, 184
fixed-action pattern 93, 174
focal vision 159
focus detector 105
fovea 128, 138, 157, 162, 184

Franceschini, Nicolas 41
Frisch, Karl von 31, 174
Furneaux, Sophie 131–40

ganglion 184
gaze-contingent masking 138
ghost crab 46
giant clam 8
glow-worm 47
Gray, John 1
Gregory, Richard 159
Grenacher, Hermann 41, 81

Habronattus (spider) 104
halteres 176
handedness (circular polarization)
    122–3
Hartline, H.K. 4
hatchet fish 66–7
Hausen, Klaus 51
Hayhoe, Mary 166
Herring, Peter 61, 67
Heteronympha (butterfly) 55
high-resolution zone, *see* acute zone
Histioteuthis (squid) 78
Homann, Heinrich 98
Hooke, Robert 27
horizon rule 96
Howard, Joe 55

image 5, 29, 47, 73
innate releaser mechanism 93, 174
interception course 38, 40
interference microscope 51, 88
inter-ommatidial angle 44, 114

Jackson, Robert 110
Jonathan I (colour-blind artist) 149
jumping spider 98–111
    courtship 99, 101
    eyes 98–104
    eye movements 101
    retina 103–4
Julesz, Bela 154

Kirschfeld, Kuno 34, 41, 43, 80
Koch, Christof 173

Layne, John 96
Lee, Dave 132
Leeuwenhoek, Antonie van 29
lens-cylinder 50, 79
light in the ocean 62–3
light-guide 31
Limulus (horseshoe crab) 50
Lingula (brachiopod) 19
Luneberg, Rudolf 25

McKinney, Travis 145
Mackworth, Norman 130
Mcleod, Peter 141
Macropipus (crab) 84
Mallock, Henry 34
mantis shrimp 111–26
    colour vision 116–17
    eye movements 113–14, 124
    eye structure 113–14
    polarization vision 118–24
Maratus (spider) 105
Marshall, Justin 115
Matthiessen, Ludwig 25
Maxwell, James Clark 25
Mennie, Neil 162
Metaphidippus (spider) 100
Metaspriggina (Cambrian fish) 16
Metazoa 184
microconsciousness 151
micro-spectrophotometer 115
microvilli 6, 119, 184
middle temporal area (MT) 156
mirror 2, 58, 63–4, 72,
    80, 85–90
mosaic theory 27
motion blur 130, 175
motor cortex 167
movement detection 8
Müller, Johannes 27, 30
multipurpose vision 173
music reading 138

Nagata, Takashi 105
Nautilus (cephalopod) 173
Nematobrachion (euphausiid) 78
neural superposition 32
Nicol, Colin 64
Nilsson, Dan-Eric 5, 19, 55, 60, 84
Noë, Alva 160

Oberwinkler, Johannes 116
ommatidium 29, 44, 115, 184
ophthalmoscope 101, 106
Oplophorus (deep-sea shrimp) 80
opsins 13–14
optical superposition eyes 32
Ordovician period 16
O'Regan, Kevin 160
'outside' memory 161

paper wasp (recognition) 174
parabolic superposition 85
Pecten (mollusc) 5
peripheral vision 135
Permian period 8–9
Phidippus (spider) 100
Phronima (amphipod) 74–5
photophore 68, 184
photopigment 155, 185
photoreceptor 6, 185
phototaxis 22
pictorial depth 155
pinhole eye 8
Platynereis (annelid) 13
polarized light 118–24
    circular 120–4
    linear 118–20
pond skater 46
Portia (spider) 110
Precambrian 11, 172
prefrontal cortex 166
protostomes 11, 185
pseudopupil 44, 74, 185

qualia 151, 176, 185
quarter-wave plate 121

quarter-wavelength film 87–90
Queen's Drive, Edinburgh 132

random-dot stereogram 155
reading 127, 137
 aloud 137
 regression 127, 137
recognition problem 92
reflecting superposition 81–3
referred pain 157
refractive index 185
 gradient 25, 50
rhabdom, fused 31, 186
rhabdomere 40, 186
rhabdomeric receptor 6, 14
rhodopsin 6, 186
Robinson, Caroline 152
resolution 33, 45
retina 3, 185
retinotopic layout 154
Rusted, Jenny 162

saccade 127–9, 142, 159, 186
saccade and fixate strategy 36
saccadic suppression 158
Salmon, Michael 97
scallop 1
scanning 108–9, 111, 124
scene network 152–3
schema 166
Schmidt corrector 3
Scopelarchus (fish) 71
semi-circular canals 130
sensory-motor contingency 161
silhouette (of fish) 65
single-chambered eyes 16
situational awareness 177
smooth pursuit 36, 144
spatial vision 22
spherical aberration 24, 186
spook fish 72
squash (sport) 145–6

stable image 157
steering 132–6
 delay 132
 geometry 136
Streetsia (amphipod) 76
superposition
 eyes 18, 47, 52, 77, 79, 186
 pupil 54
Syritta (hoverfly) 35

tangent point (on road) 133, 135
task-related objects 165
Tatler, Ben 131, 159
tea-making (gaze analysis) 162–5
temporal pattern (claw waving) 97
top-down control 162
total internal reflection 186
transduction cascade 13
Tridacna (giant clam) 8
trilobites 9, 16
Trumper, Victor 141
tubular eyes 72

vestibulo-ocular reflex 130
visual cortex 150, 167–8, 187
Vogt, Klaus 81

Walls, Gordon 130
waveguide modes 59
wavelength 149
Weaver, Homer 139
Westheimer, Gerald 10, 101

X-ray telescope 84

Yarbus, Alfred 162–3
Young, Thomas 24

zebra spider 100
Zeki, Semir 149
zombie 170, 177
Zurek, Daniel 104

THE GOLDILOCKS PLANET

The 4 billion year story of Earth's climate

Jan Zalasiewicz and Mark Williams

JAN ZALASIEWICZ & MARK WILLIAMS

978-0-19-968350-5 | Paperback | £10.99

"A balanced, well written, mostly comprehensive and well-argued book."

Times Higher Education Supplement

In this remarkable new work, Jan Zalasiewicz and Mark Williams demonstrate how the Earth's climate has continuously altered over its 4.5 billion-year history. The story can be read from clues preserved in the Earth's strata—the evidence is abundant, though always incomplete, and also often baffling, puzzling, infuriating, tantalizing and seemingly contradictory. Geologists, though, are becoming ever more ingenious at interrogating this evidence, and the story of the Earth's climate is now being reconstructed in ever-greater detail—maybe even providing us with clues to the future of contemporary climate change.

WHAT IS LIFE?

How Chemistry Becomes Biology

Addy Pross

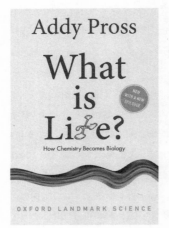

978-0-19-878479-1 | Paperback | £9.99

"Pross does an excellent job of succinctly conveying the difficulty in crafting an unambiguous general definition of life and provides a road map to much of the work on the origin of life done by chemists in the past 50 years. The book is worth the read for these discussions alone." ***Chemical Heritage***

Living things are hugely complex and have unique properties, such as self-maintenance and apparently purposeful behaviour which we do not see in inert matter. So how does chemistry give rise to biology? What could have led the first replicating molecules up such a path? Now, developments in the emerging field of 'systems chemistry' are unlocking the problem. The gulf between biology and the physical sciences is finally becoming bridged.